TACKLING GEOGRAPHY COURSEWORK

John Pallister & Ann Bowen

Hodder Murray

A MEMBER OF THE HODDER HEADLINE GROUP

Map on p 87 reproduced by permission of Ordnance Survey on behalf of The Controller of Her Majesty's Stationery Office, © Crown Copyright 100036470.

Photographs on pages 15, 20t, 20b, 67t and 68 © Education Photos/John Walmsley. All others © John Pallister.

Although every effort has been made to ensure that website addresses are correct at time of going to press, Hodder Murray cannot be held responsible for the content of any website mentioned in this book. It is sometimes possible to find a relocated web page by typing in the address of the home page for a website in the URL window of your browser.

Orders: please contact Bookpoint Ltd, 130 Milton Park, Abingdon, Oxon OX14 4SB. Telephone: (44) 01235 827720. Fax: (44) 01235 400454. Lines are open from 9.00–6.00, Monday to Saturday, with a 24-hour message answering service. Visit our website at www.hoddereducation.co.uk

First published in 2005 by
Hodder Murray, a member of the Hodder Headline Group
338 Euston Road
London NW1 3BH

Impression number   10 9 8 7 6 5 4 3 2
Year   2010 2009 2008 2007 2006

Typeset in GillSans by Fakenham Photosetting Ltd, Fakenham, Norfolk.
Printed in Italy

A catalogue record for this title is available from the British Library

Student's Book: ISBN 0 340 88676 5
                EAN 9 780 3408 8676 2

Teacher's Book: ISBN 0 340 88675 7
                EAN 9 780 3408 8675 5

Contents

Chapter 1
Introduction

Coursework in geography – what is it?

Written examinations have set dates and times. An examiner, who you do not know, has set questions that you have never seen before. You have a fixed length of time in which to answer the questions. You need to perform well in the examination in order to obtain a good grade, even if you are not feeling at your best on the day.

Coursework is different.

■ There are no fixed times and dates for doing it. Coursework can be done at any time during the course, provided that it is completed before the date fixed for handing it in.

■ There is no examiner to tell you what you should investigate. You or your teacher chooses the topic, not people from the Examination Board unknown to both of you.

■ There is more time to do the work and write it up. This allows you to discover geographical information from **fieldwork**. You have time to practise geographical skills like drawing maps, diagrams and graphs, either by hand or with the help of a computer.

A written examination is like a sprint. It is over quickly after a long period of preparation. Coursework is more like a long distance race or mini-marathon. You can pace yourself and then speed up towards the finish to complete the writing up.

Fieldwork

This is data collected outside the classroom. You go out of school to discover geographical information, either working by yourself or as part of a group. It can be much more fun than sitting in classrooms, especially in good weather. The big difference is that you are finding out your own geographical information. Observing, counting, measuring and discovering about places, that is what geography is all about.

For example, river fieldwork is usually popular with geographers (Figure 1.01). Water is attractive to people of all ages, but you must always remember that undertaking fieldwork in rivers can be dangerous. Why is the river shown in Figure 1.01 a suitable river for undertaking fieldwork measurements?

DOs AND DON'Ts

DO

✔ Make a timetable of work

✔ Stick to it ruthlessly

DON'T

✘ Dither and delay doing the work

✘ Make doing the coursework never ending

Figure 1.01 River fieldwork

How to use this book

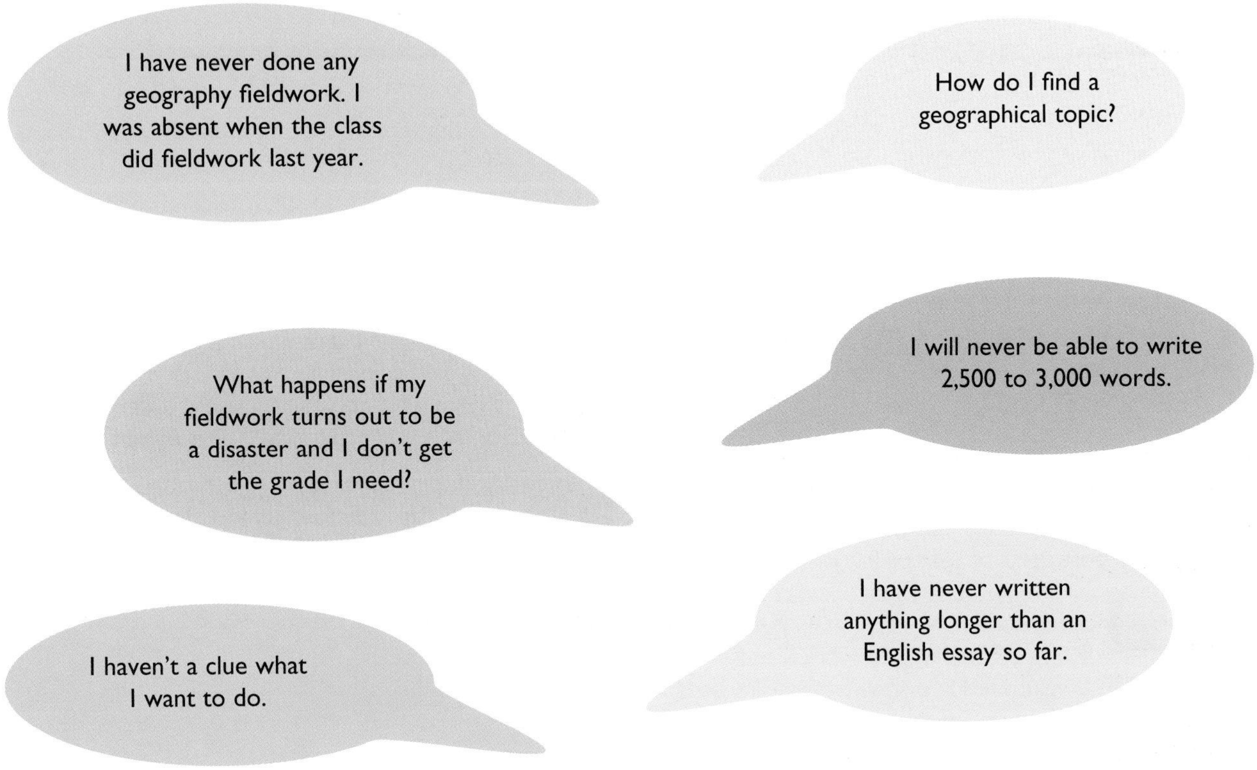

I have never done any geography fieldwork. I was absent when the class did fieldwork last year.

How do I find a geographical topic?

What happens if my fieldwork turns out to be a disaster and I don't get the grade I need?

I will never be able to write 2,500 to 3,000 words.

I haven't a clue what I want to do.

I have never written anything longer than an English essay so far.

Figure 1.02 Student worries before starting coursework

The majority of geography students wobble and feel weak at the knees when coursework is first mentioned – almost like a heap of jelly. The task can seem formidable. At first virtually everyone is clueless about what is expected (Figure 1.02).

The message is don't panic. Stay calm. Work your way through this book. Concentrate on those parts that are relevant to the geographical topic that you decide to choose. The first four chapters in this book take you through the coursework processes. They make a stage-by-stage guide to coursework.

■ There are plenty of ideas in **Chapter 2** about topics in geography. You should be able to find one that interests you.

■ Information about ways to collect fieldwork data is given in **Chapter 3**. If you don't feel comfortable stopping people and asking questions, you can always choose fieldwork that depends more upon observing, counting or measuring.

■ In **Chapter 4** methods for presenting and processing data are described. If you thought there were only maps, graphs and pie charts, think again. You will be introduced to more than 20 methods from which to display your data and add variety to your final work.

■ Writing up the coursework is often the hardest part. In **Chapter 5** you will find guidance on what to include, as well as hints about how to end on a high note.

■ **Chapter 6** contains examples of the geographical topics most commonly used in coursework. You will find:
  – More suggestions for titles
  – Further details about methods of data collection
  – Student coursework plans
  – Examples of students' work.

  This chapter demonstrates how to put together complete pieces of coursework. Even if rivers is not your topic, you are advised to look at the examples of students' work on pages 70–73, which are of grade A* standard. They show you what to aim for.

■ Don't ignore **In conclusion**, the short final chapter. It contains helpful information about how to please the people who will mark and moderate your work.

■ Finally, if you come across geographical words or terms that you do not understand, look in the **Glossary**. Remember that the **Index** is the book's 'search engine'.

Coursework structure

Schools and colleges organise coursework in many different ways, but there are two main methods, as described below.

Individual studies

Each student chooses his/her own topic. He/she may be the only person in the class who is doing that topic.

1 Look at the flowchart on pages 8–9 so that you learn about what needs to be done.

2 Find a geographical topic that interests you (Chapter 2). Discuss it with your teacher to find out whether it is going to be possible to do it in your local area.

3 Keep your teacher informed about what you have done and what you intend to do next.

Teacher-led and group studies

All members of the class investigate the same geographical topic. Some, most or all of data collection by fieldwork is organised and guided by the teacher. Typically, data collection is undertaken by two, four or six people working together. This can be a good way to work, because a lot more data can be gathered in a shorter time. In many river and coastal studies several people need to be involved because it is often necessary to use and manoeuvre equipment, take measurements and note down the results all at the same time. In CBD studies, by dotting small groups around the city centre, it is possible for traffic or pedestrian counts to be carried out simultaneously, which allows accurate comparisons to be made.

TOP TIPS

Your teacher is . . .

- The No. 1 resource for coursework
- The person who knows best what is needed
- The person who understands how the work will be marked
- The No. 1 source for guidance and help.

Although everyone in the group will have the same data, the work in the next stages of the coursework operation (processing and presenting the data and writing it up) must be done on an individual basis. Each student must work alone. This is a very important examination requirement.

If you are hoping to obtain a high grade for your coursework, it is vital that you include something that is all your own work – something that makes your work different from (and hopefully better than) the work produced by other members of the class or group. Marks are awarded for what is called *individual initiative*. You must make sure that you claim these marks. You claim them by extending the group work. The box below shows the three main options open to you to add some individual input to your coursework.

Although the main aim may be to make your work different, there are other benefits. The more data you collect, the more there is to write about and the more you will have for presenting as maps, diagrams and graphs.

OPTIONS FOR EXTENDING GROUP WORK

1 Collect the same type of data, but in a **different place**.

Advantages:

- You know how to collect the data.
- You can compare two places.
- There is more to write about when looking for similarities and differences.

2 Go back to the same place, but collect **more of the same type of data OR a new type of data**.

Advantages:

- You know how to collect the data.
- You can collect data at different times or on different days (which can be compared with group collected data).
- Writing opportunities are greater when you are looking for similarities and differences.

3 Go to a **different place** and collect a **new type of data**.

Advantages:

- More of the data will be unique – different from that collected by all the others.
- The work will look more like an individual study.
- There is more chance of using a greater variety of methods of presentation.

Figure 1.03 What characteristics typical of a CBD can be seen on this photograph? Look at the example below for suggestions about extending a group study of a CBD.

EXAMPLE – HOW TO EXTEND A CBD STUDY

Title of work – 'What are the characteristics of the CBD in City X?'

Data already collected by the group:

- Survey of land uses (at ground level, e.g. shops, banks, offices)
- Traffic surveys in and around the edge
- Parking survey (car parks and parking restrictions).

Possibilities for new data collection to extend the work:

- Pedestrian counts
- Questionnaire for shoppers
- Another traffic survey but at a different time (e.g. rush hours)
- Another parking survey but on a different day (e.g. on a Saturday).

How to use this book

When you are ready to think about extending your coursework, do the following:

1 Look in the index to find the page numbers for your topic. If you decide to search for another topic that you can't find in the index, Chapter 2 will help.

2 If you need ideas about new methods to collect data, look at Chapter 3.

3 Add some variety to your data presentation by seeking ideas for new methods from Chapter 4. Don't get stuck in the tramlines by drawing only bar graphs and pie charts.

4 Look in Chapter 5 for advice on writing up the work.

Coursework for a geography examination

STAGE 1 – CHOOSING A GEOGRAPHICAL TOPIC

- Think of an idea – preferably local and with easy access.
- Talk to your teacher about it.
- Read up on the geography of the topic chosen.
- Agree a title.
- Write down what you want to find out – these are your aims.
- Consider the data you need to collect and how you will collect it.

Read Chapter 2 to help you.

DOs AND DON'Ts

Your topic must be **geographical** and it must be about **places**.

DON'T

- ✗ Compare the prices of foods in different supermarkets

- ✗ Give an Internet guide to shopping

- ✗ Describe in detail the layout and inside features of shops, leisure centres, etc.

- ✗ Write about the different types of cars in a car park or traffic survey

STAGE 2 – COLLECTING THE DATA

- Collect as much data as you can in the time available.
- Some must be collected by fieldwork.
- Think about safety.
- Some equipment might be needed.
- Look for other sources of geographical information such as maps, books, newspapers and the Internet.

Read Chapter 3 to help you.

STAGE 3 – PRESENTING AND PROCESSING THE DATA COLLECTED

- Show the location of the study area on a map.
- Put any statistics you have collected in tables.
- Draw a variety of graphs.
- Present some of the data as maps.
- Illustrate your work with labelled sketches and photographs.

Read Chapter 4 to help you.

STAGE 4 – WRITING IT UP

- Write about your maps, tables, graphs and diagrams.
- Explain what they show.
- How do they help to meet the aims of the enquiry stated in your *Introduction*?
- Try to give an overall conclusion.
- Think about the geographical significance of your work.
- Evaluate the strengths and weaknesses of what you have done.
- With more time, what further work could have been done?

Read Chapter 5 to help you.

How will your work be marked?

A short summary of what will gain you marks is given below.

Planning and data collection
Marks will be given for:

- Making a clear statement of your aims
- Referring to the geographical background of the topic
- Using different methods of data collection
- The amount of data collected.

This section is worth about 40% of the total marks.

Data presentation
Marks will be given for:

- Using a range of different methods of presentation (i.e. several different methods appropriate for the data to be shown)
- Presenting the data accurately
- Adding titles, keys and labels to maps, graphs, sketches and photographs.

This section is worth about 20% of the total marks.

Analysis, interpretation and conclusion
Marks will be given for:

- Describing the results from the data collected and presented
- Suggesting reasons for your results
- Making links between different types of data
- Drawing your conclusions
- Relating your conclusions to the original aims
- Attempting an overall evaluation of the work done.

This section is worth about 40% of the marks.

TOP TIPS

- Use as many different methods of presentation as you can.
- Highlight the most important features of diagrams and maps.
- Add labels to sketches and photographs.
- Make sure you add titles and keys.

TOP TIPS

- The written part is worth a lot of marks.
- Analysing the data is a higher order skill than presenting it; that is why it is worth more marks.
- Look back at your aims and say how well they have been achieved.
- Don't be afraid to mention any aims not achieved and any difficulties you may have had.

Using ICT

The chances are that the majority of you will choose to use computers and information technology at some point in the coursework process without any encouragement to do so by your teacher or by Examination Board regulations.

All the exam boards require some use of ICT (Information and Communications Technology) before full marks for coursework can be obtained. What percentage of the marks is reserved for showing ICT proficiency? It is often difficult to get a direct answer to this question from Examination Boards, but it seems to be in the order of 5–10% of the total marks.

Think of the significance of this:

■ **90–95%** of the marks are for **geography** and demonstrating **geographical skills**.

■ **5–10%** of the marks are for demonstrating **ICT skills**.

In other words, the marks reserved for ICT are worth having, particularly if you are seeking the highest grade, but it is the geographical worth of the coursework that is the main determinant of mark and grade. This should encourage the minority among you who are unhappy and unwilling participants in the ICT revolution. Equally it should be a warning to the computer enthusiasts among you that brilliant ICT skills do not automatically translate into a high grade in geography in the way that they would in a computer studies examination.

In the context of geographical coursework, the following represent the major opportunities for ICT use.

Collecting secondary data

Once upon a time, not so long ago, the only way to collect secondary data was by visiting the reference library to use encyclopaedias, government publications such as the census, and statistics from international organisations like the UN (United Nations) – provided that you could visit during library hours! Now everyone has access to all this data, and much more, at school or in the home, from CD-Roms and the Internet, etc, 24 hours a day, 365 days a year (and for 366 days in leap years!).

Presenting data

Using computer programs saves a lot of time compared with producing tables, charts, graphs, diagrams and maps by hand. It also gives results that are visually pleasing and of high quality – sometimes of book quality. Because the task of presenting data is so greatly speeded up, more time is released for writing up and analysing your results, which is usually worth more marks.

■ Spreadsheets

Spreadsheets can store, process and present large quantities of data, such as all your questionnaire results. Once the manual task of transferring the results onto the spreadsheet has been completed, the computer will, under instruction, do calculations from the figures and show them in tables and graphs. For example, from the results of a shopping questionnaire given to different age groups, the shopping habits of each age group can be shown separately. You can find out where people of that age group shop, what they buy and how much they spend. The results can be compared between the different age groups. The patterns of shopping identified can then be explained in your *Analysis* section.

TOP TIP

Check with your teacher that your specification or syllabus does not require a certain number of hand-drawn examples to be included for full marks.

Mapping data

A mapping program can be used to show patterns; for example, you can plot different types of shops on a base map of a town. Once a pattern is shown, you can suggest geographical explanations for it. Mapping programs are also ideal for showing changes; for example, land uses on a farm over several years can be compared. Scanning printed maps is possible and, once on screen, you can add information relevant to your study.

Graphs and charts

Certain types of graphs generated by computer programs would be far more complicated to draw by hand. For example, some graphs are given 3-D effects to improve appearance, such as vertical bars that are made to look like building blocks and pie graphs that have raised edges and a hole in the centre (the 'doughnut' reference on Figure 1.04). When using these effects always check that the data is being displayed accurately. For example, with bars that are made to taper like a pyramid, values are shown by the relative heights of the tops of the pyramids (i.e. by pyramid length) and not overall size.

Showing and annotating photographs

Images from digital cameras allow you to select what is best for illustrating your work; you can concentrate on and highlight those parts of the picture that are of greatest significance. If you have the use of a good scanner, photographs from a variety of published sources can be inserted. These can be placed in the most relevant places in the written text to give them maximum impact.

Writing the text

The most widespread use of ICT is for word processing the text. Not only does printed text give a higher quality presentation than writing it out by hand, but you are free to draft and re-draft what you have written as many times as you like. You can keep making improvements either by changing what is there or by extending it. Also you can vary font size and style for added impact. The spell-check facility should ensure greater accuracy in spelling (although not all geographical terms are in the computer's dictionary). Using the computer offers great benefits, but beware of becoming a machine yourself by no longer thinking about what you are doing.

Summary

The possible uses of ICT are summarised for you in Figure 1.04. The most common uses of computers by students when completing their geography coursework are listed in the centre circle. In the segments around the edges a variety of other possible uses is shown, some of which require computer skills of a higher order. Variety is a good way of demonstrating competence in ICT skills.

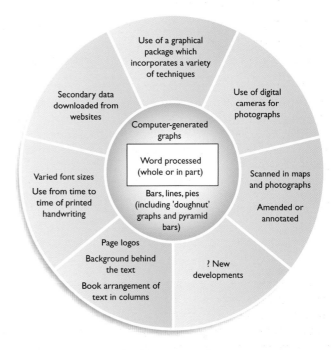

Figure 1.04 Uses of ICT in geography coursework

DOs AND DON'Ts

DO

✔ Keep thinking and enquiring: 'Why am I using the computer? Is it helping my study?'

✔ Keep saving your work in case of loss of power

✔ Back up your work regularly

DON'T

✘ Repeat the same type of graph time and time again

✘ Restrict yourself only to the methods of presentation that are included on the computer program

✘ Believe that everything must be right if done on a computer (remember – people feed computers)

Chapter 2
Choosing a topic

Ideas within topics

Figure 2.01 shows just a few of the many possible ideas within geography that you could study for your coursework. Some of the suggestions are physical; others are human. Opportunities for some are greatest in towns and cities; others can be better studied in villages and rural areas.

Although there are many and varied possibilities, in practice your choice is likely to be limited by:

- Personal likes and dislikes
- Where you live or places you visit often
- The need for safety while doing fieldwork.

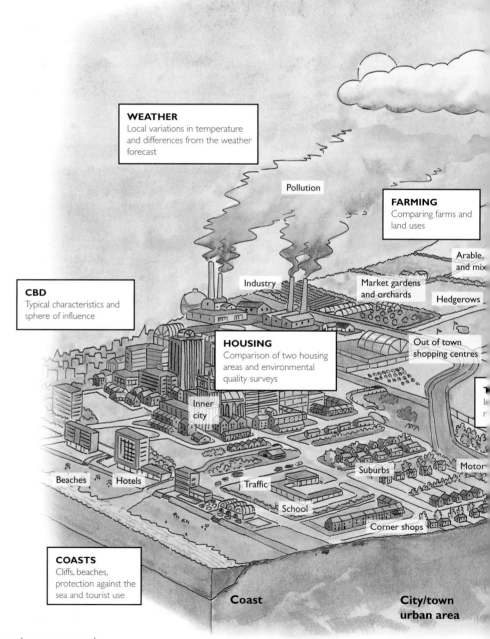

WEATHER
Local variations in temperature and differences from the weather forecast

Pollution

FARMING
Comparing farms and land uses

Arable, and mix

Industry

Market gardens and orchards

Hedgerows

CBD
Typical characteristics and sphere of influence

HOUSING
Comparison of two housing areas and environmental quality surveys

Out of town shopping centres

Inner city

Suburbs

Motor

Beaches

Hotels

Traffic

School

Corner shops

COASTS
Cliffs, beaches, protection against the sea and tourist use

Coast

City/town urban area

Figure 2.01 Ideas for study for geography coursework

1 List the topic headings from Figure 2.01 which it would be possible to study:

(a) Within walking distance of home or school

(b) Within 20 miles of your home.

2 (a) Which of the topic areas given do you enjoy most?

(b) Describe what can be studied for coursework in your favourite topic.

Concentrate your study of Figure 2.01 upon those parts that match where you live or visit often.

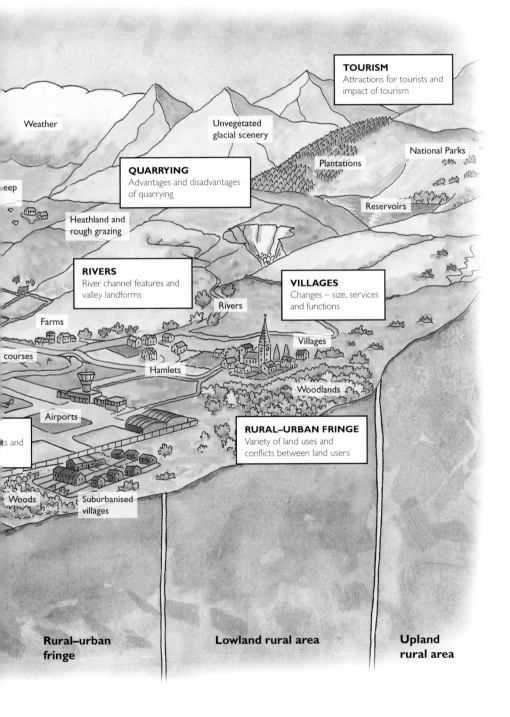

TOURISM
Attractions for tourists and impact of tourism

Weather

Unvegetated glacial scenery

National Parks

Plantations

QUARRYING
Advantages and disadvantages of quarrying

eep

Reservoirs

Heathland and rough grazing

RIVERS
River channel features and valley landforms

VILLAGES
Changes – size, services and functions

Rivers

Farms

Villages

courses

Hamlets

Woodlands

Airports

RURAL–URBAN FRINGE
Variety of land uses and conflicts between land users

ts and

Woods

Suburbanised villages

Rural–urban fringe

Lowland rural area

Upland rural area

How to find a topic

Think about what interests you in geography

Copy and complete the Topics Questionnaire below.

TOPICS QUESTIONNAIRE

	Like	Maybe	Dislike
Physical geography			
Rivers	O	O	O
Coasts	O	O	O
Vegetation	O	O	O
Soils	O	O	O
Weather	O	O	O
Human geography			
Urban settlement – cities and towns	O	O	O
Rural settlement – villages	O	O	O
Farming	O	O	O
Industry	O	O	O
Retail/shopping	O	O	O
Leisure/recreation	O	O	O
Tourism	O	O	O
Transport	O	O	O
Environmental and local issues			
Pollution	O	O	O
Quality of landscape	O	O	O
Environmental impact of quarrying/ opencast mining	O	O	O
Plans for change – a new route for a road	O	O	O
Planning an out-of-town shopping centre	O	O	O
Plans for a new housing estate in the village	O	O	O
Others			
Add any other possible topic areas to the list.			

TOP TIPS

If you still need help finding a suitable topic, you could try the following.

- Make a list of topics that simply interest you.

- Look at the local press. Is there an issue that is being given a lot of publicity at the moment? For example: new by-pass, new out-of-town shopping centre, new housing estate, pollution of the environment, controversy over the location of a new mobile phone mast. It is much better doing coursework with a local flavour to it.

- Make a list of your preferred methods of data collection: questionnaires, measurements, observation, websites, published materials (e.g. maps, books and newspaper reports). Your choice could affect the topic.

- Do some research to see what is possible.

- Ask your teacher to suggest examples of good local topics. Teachers are useful in helping to find a title that matches your aims.

- Don't go to your teacher with the brain tank on empty. The more you have already thought about it, the better the suggestions your teacher will be able to make.

Think about which data collection methods you prefer to use

MINI QUESTIONNAIRE

Am I a questionnaire type?

	Yes	Maybe	No
Would I like stopping people in the street and asking them questions?	O	O	O
Would I like calling at houses of people I didn't know with questionnaires?	O	O	O
Would I be happy giving out questionnaires to people I know in the area where I live?	O	O	O

MINI QUESTIONNAIRE

Would I like using equipment to take measurements, such as in a river, on a beach, or across a valley?

	Yes	Maybe	No
Do I enjoy using equipment?	O	O	O
Would I enjoy taking measurements out of doors in bad weather?	O	O	O
Do I like all my measurements to be accurate?	O	O	O

MINI QUESTIONNAIRE

Which of these methods would I be happy to use?

	Yes	Maybe	No
Doing traffic counts	O	O	O
Counting numbers of people	O	O	O
Observing (e.g. shops, houses)	O	O	O
Sketching (e.g. buildings, countryside)	O	O	O
Searching the Internet/reference materials	O	O	O

It is important to choose methods of data collection that suit you. There is no point choosing a topic that relies upon answers from questionnaires if you feel uncomfortable stopping people you don't know and asking them questions (see Figure 2.02). That is not going to lead to successful coursework. There are plenty of other topics for which all the fieldwork can be completed by observation. This can be supported by searching for reference materials on the Internet, in libraries or from newspapers. Work through the mini questionnaires (above) to ask yourself some starter questions.

Figure 2.02 School students interviewing local residents. Would you be happy doing this?

Depending on your answers, you may be in a strong position and know what you want to study, or in a weaker position and still need time to decide. Most probably, though, you will be in a position somewhere between the two. Look at the Top Tips box on page 14.

Think about location

There are good reasons why you should do fieldwork in the area close to your home.

- You are free to go out and collect data whenever it suits you.
- You can choose the best times for doing the work, for instance on a weekend or an evening in summer.
- If, on the visit, there isn't the time to collect enough data, you can go back and collect more.
- There are more opportunities for collecting data at different times or under different conditions.

In other words, repeat visits are possible in a way that they are not if the work is done while you are away on holiday. Also it is easier to check out that you have access to sites in your study area. Florida may sound more exotic than Falmouth, Forfar or Fulham, but it is a better site for Disneyworld than it is for your coursework! Don't consider using an area where you go on holiday for your coursework, unless you spend a lot of time there (at least one month each year) so that you know the area really well and understand the coursework possibilities.

Consider safety and security as well. In some cases, both for your own safety and for accuracy of measurement, it may be better if you work as part of a small group. Depending on the type of fieldwork you are doing, between two and six people is likely to be best. Group collection of data is allowed by all the Examination Boards, but *presenting and writing up the data must be all your own work*. This is an important rule.

Finding a title

Your title needs to describe what you are trying to find out in your coursework. The best titles are often written as a hypothesis, a question, an issue or a problem (see Figure 2.03). Remember that what you study must be geographical (see Figure 2.04).

A **hypothesis** is an idea that can be tested. It usually expresses a relationship between two or more variables. In testing a hypothesis you need to produce evidence to support or dismiss or amend the hypothesis. Remember the purpose is not simply to prove whether or not it is correct. Examples of hypotheses are: 'The temperature across the valley will vary with height and aspect' and 'Some shops and services cluster more than others'.

A **question** often has one or more of the following key words -- What? How? When? Where? Why? ... and always has a question mark at the end.

An **issue**-based study considers an argument. There are often two or more sides to the issue or conflict and you need to make a judgement about the different viewpoints. One example of an issue-based study would be 'What are the conflicts of land use in the Malham Cove area of the Yorkshire Dales National Park?'

A **problem-solving** title is often a question. An example would be: 'What would be the best location for a by-pass in Durham City?' You need to research the problem and suggest possible solutions supported by explanations.

COURSEWORK

A question is . . .

An issue-based study . . .

A hypothesis is . . .

A problem-solving title . . .

Figure 2.03 Types of title

Places

Population

Physical features and processes

Tourism

Transport

Geographers:

- Make up-to-date descriptions of landscapes, both physical and human, and they try to explain the patterns they find
- Study the way people change and interact with the landscape
- Look at the challenges facing people and the choices people can make about the environment.

Weather

Farming

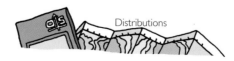
Distributions

Settlements

Figure 2.04 What is geographical?

Examples of titles

1 Rivers (see pages 66–73)
- How and why does the channel of River X vary downstream?
- What changes in the depth and flow of River X occur during the year?
- Hypothesis – that the size of particles on the river bed increases upstream.

2 Coasts (see pages 74–77)
- What are the main features of the beach and cliffs at X?
- What are the similarities and differences between the coasts at X and Y?
- How successful are the methods used to protect the coastline at X?
- Why is the beach at X more important for tourism than the beach at Y?

3 Urban (see pages 78–87)
- In what ways is the CBD different from the other urban zones in Town/City X?
- Where does the CBD of City X begin and end?
- How and why do land uses change between the centre and edge of Town X?
- What is the sphere of influence of the out-of-town shopping centre at X?
- Hypothesis – that the environmental quality of houses and housing areas increases away from the centre of Town X.
- Why do house prices vary so much within City X?
- To what extent does the layout of Town X fit the Burgess model?

4 Leisure and tourism (see pages 88–91)
- What attracts tourists/visitors to X?
- What is the sphere of influence of the tourist attraction/leisure facility at X?
- What is the impact of visitors in the National Park/Country Park/Local Park?
- Hypothesis – that the numbers of visitors and their impact upon local residents increases during school holidays.
- Where would be the best place to build new car parks and picnic sites for the increased number of visitors?

5 Vegetation and soils (see pages 92–93)
- How and why do vegetation and soils change across the valley of River X?
- Does a relationship exist between plant type, slope angle and soil characteristics?

6 Weather (see pages 94–95)
- How accurate are the national weather forecasts in relation to the weather recorded at the school weather station?
- An investigation into the microclimate of a large park.

7 Rural areas (see pages 96–99)
- What are the differences between Villages X and Y with respect to layout and growth?
- Does Town/Village X need a by-pass?
- In what ways and why has the village of X changed in the last 20 years?
- To what extent is Village X a commuter settlement?
- In what ways and why are the services available different in villages in the X valley?
- What are the similarities and differences between two or more farms?
- What is the impact of people buying second homes in a tourist area?

8 Environmental pollution (see pages 100–101)
- Why does the amount of litter pollution vary from place to place?
- Is our local stream polluted?
- Hypothesis – that visual pollution in urban areas increases towards the centre.
- Should the company be allowed to increase the size of its quarry in the National Park?

9 Transport (see pages 102–103)
- What is the impact of the local airport/bus station on the people living in the area around it?
- Why do traffic flows in X vary greatly at different times of the day?
- What are the traffic problems in X and how can they be reduced?
- Where should the new by-pass be built?

Checklist

Before you finally settle on your coursework title, ask yourself the following questions.

- ○ *Is my title geographical?* (check with Figure 2.04)
- ○ **Will the choice of title allow me to collect a variety of data?**
- ○ *Is my choice of title too broad?* (e.g. what factors affect the location of industry in Wales?)
- ○ *Is my choice too narrow?* (e.g. what are the houses like in my street?)
- ○ **Has my teacher approved the title?**

TASKS

1. (a) Describe what is meant by 'geographical'.
   (b) Explain the difference between physical and human geography.
   (c) Look back at the DOs and DON'Ts box on page 8. Give reasons why the three examples stated there are non-geographical.

2. (a) From the titles given on this page, give examples of the four types of title named in Figure 2.03.
   (b) What is the most common type of title used? Suggest why.

3. (a) How many of the nine topics listed on this page would it be possible to study in your local area?
   (b) For which topic is there most choice? Explain.

Planning your coursework

Write down your idea

Write down your chosen topic first, for example, 'A river study'. Then make an attempt at a title, for example, 'Is River X a typical river?'

Think about the practical possibilities

Be realistic. Ask yourself: 'Is it going to be possible to put my idea into practice?'

- **Where**. How far from home is the nearest river?
- **Size**. Is the river the right size for taking fieldwork measurements, neither too large nor too small?
- **Access**. Can suitable sites for doing the fieldwork be reached?
- **Equipment**. Do you have all the equipment and clothing needed?
- **Safety**. Will you be able to do the fieldwork alone or will you need others to help?

Plan what you are going to do

What aims or hypotheses can I have?
Examples of hypotheses for a river study:

- The river channel will become wider and deeper downstream.
- The river's discharge and speed of flow will increase downstream.
- The gradient of the river will become less steep downstream.
- The size of boulders in the river's channel will decrease downstream.
- The shape of its valley will become wider and less steep downstream.
- There will be more human uses of the river and its valley downstream.

What methods of data collection can I use?
Examples of methods used in river studies:

- Taking measurements of channel width, channel depth, water's speed of flow, length of boulders on the bed, channel gradient, slope angles on the valley sides.
- Observation of channel and bank features, including the use of field sketches and photographs; observation of human features.
- Questionnaire to users of the river and its banks.

Figure 2.05 The River Tees – an unsuitable river for study. Why?

A. My idea for a geography topic

IDEA   Studying an out-of-town shopping area.

TITLE   Has it got the typical characteristics of an out-of-town shopping area?

B. Putting my idea into practice

STUDY AREA   The Arnison Centre outside of Durham.

ACCESS   I can walk to it. There are also buses. There is open access everywhere, although I will need permission from managers if I am going to stop shoppers as they go into the shops.

EQUIPMENT   I will make 80 copies of my questionnaire for shoppers and keep them in a plastic folder. I have my own automatic camera. I don't think I need anything else.

SAFETY CHECK   A friend has agreed to come along with me the first time I give out the questionnaires. It is an open area with lots of people around, so I should be alright by myself.

C. Planning what I am going to do

AIM/QUESTIONS/HYPOTHESES   What is its sphere of influence? What shops and services are available? Which shop is the most popular? How do most people travel to the centre? Is it easily accessible? What are people's views on out-of-town shopping areas? This is all I can think of at the moment, but I may be able to think of more before I start work.

METHODS OF DATA COLLECTION

OBSERVATION   Land use survey of names of shops and services and the types of goods sold. Environmental quality survey.

QUESTIONNAIRES   I am going to hand these out to shoppers on two different days of the week. I will ask about 10 questions, like: Where did you travel from? By which means of transport did you travel? How often do you visit? Which shop do you most regularly visit? Are you in favour of out-of-town shopping areas? This will be my main way of getting data.

MEASUREMENT   None

OTHERS   Letter to the Council about reasons for building the Arnison Centre. Looking in the reference library for information.

Figure 2.06 An example of a completed planning sheet. It was filled in by a student preparing to study an out-of-town shopping centre. It shows you what to do. Always do some research to make sure that you are going to be able to collect enough data. Remember that good data collection helps both presentation and analysis.

Chapter 3
Collecting data

Types of data

The data are pieces of information you will collect in order to answer the aims of your enquiry. There are two main types of data, which are explained and illustrated below.

Primary data

These are 'new' pieces of information collected by you in the field. All enquiries for use in examinations must contain some primary data collected by fieldwork. Safety when undertaking fieldwork is of great importance.

Primary data include what you:

- Observe (by writing notes, taking photographs, drawing sketches and making assessments of environmental quality)
- Measure (e.g. river speed or widths of shop fronts)
- Count (e.g. pedestrians, cars)
- Find out by asking people (e.g. by questionnaires and interviews).

Figure 3.01 Collecting primary data

Secondary data

These are 'old' pieces of information obtained from other sources. Someone else has collected the data you are going to use.

Secondary data include:

- Any information taken from websites
- Information from libraries, travel agents and estate agents
- Weather information obtained from TV and newspaper reports
- Everything taken from books and newspapers
- Information from local councils and government sources.

It is a good idea to keep a list of all the secondary sources used so that you can name them in the list of references at the end of the coursework.

Figure 3.02 Collecting secondary data

Planning your data collection

Carefully plan:

- **What** data to collect
- **Where** you will collect the data
- **When** you will collect it
- **How** you will collect it
- What **materials** and **equipment** you will need.

It is a good idea to write yourself a plan or diary to show what you are going to do and when.

Full details about what, where, when, how and which equipment should be used will be needed as part of your *Introduction* and in your chapter about *data collection* (see page 58. To help you explain your methods of data collection, you may include photographs and sketches as illustrations of the methods used. In some cases a list of equipment used will also be helpful. One empty copy of any record sheets and questionnaires must be included here as well. All of these provide evidence of the fieldwork.

Timing your data collection

In certain types of coursework, *when* you undertake your data collection is important. Examples include:

- How something changes over time, such as the number of tourists or shoppers, river flow and the weather
- Studies of physical features such as rivers, beaches and vegetation, and human topics such as land uses on farms, which can be different between summer and winter
- Transport or pedestrian counts – do you need to count at busy times, slack times or a mixture of the two?

TOP TIPS

An enquiry that is given top marks for data collection has the following:

- Primary data collection (fieldwork)
- A variety of methods of data collection
- Accurate data collection
- Appropriate data collection (for the title)
- Sufficient data collection to answer the aims of the coursework
- If used, secondary data which support the aims.

SAFETY – when doing fieldwork

Wear suitable outdoor clothing.

Tell an adult where you are going and what you are doing.

Work in pairs or small groups when necessary.

Also say approximately what time you will return.

DOs AND DON'Ts

DO

- ✔ Collect appropriate data for your aims
- ✔ Be as accurate as possible
- ✔ Follow the Countryside Code
- ✔ Make a note of any problems – you will need to evaluate the success of your data collection

DON'T

- ✗ Leave it until winter to take the photographs
- ✗ Collect too little or too much data
- ✗ Trespass or break the law
- ✗ Do anything unsafe or dangerous

Sampling

If you observe *every* shop in the CBD, hand out questionnaires to *every* person visiting the tourist attraction, or count *all* the traffic on a road during the day, you are recording what is known as the **total population**. This term doesn't just refer to people; rather it covers every single thing or person that you could have investigated in your fieldwork.

In most fieldwork an investigation that studies the total population is impossible. How can you hope to measure the size and shape of every pebble in a river channel, or have the time to ask questions about holiday destinations to everyone who passes through an airport in one day? Instead you need to take a **sample**. This is where you confine your investigation to a selection from the total. The hope is that what you learn from your sample investigation of a few will turn out to be a faithful reflection of the full picture that you do not have the time and resources to study.

How you choose the sample to be investigated is of critical importance.

■ You must try to eliminate human bias. When undertaking questionnaires in the street there is a temptation to ask only those who look as though they will answer your questions – or you select good-looking and cool males and females only!

■ You must try to make your sample as representative of the total population as you can – all different types, from all areas, need to be included to increase the confidence that your partial results will be a reliable reflection of the total.

Sampling methods

There are three recognised methods of sampling. You should choose the one that best fits your type of study and your fieldwork opportunities.

1 Random sampling

The people or places for study are selected completely by chance. If you use random numbers, every person or place has an equal chance of being the one selected for study.

Method

The best is to use tables of random numbers (which can be computer generated); examples of these are shown in Figure 3.03. As arranged, they give you random values from 0–99. You do not have to begin with the top left number of 84; instead you could begin in the bottom right corner with 34, and work backwards. You can start anywhere and go in any direction, provided that you keep going in the same direction throughout. If you need numbers above 100 you can select numbers from the table

84	51	03	65	79	00	03	56	58	21
96	29	15	04	82	91	41	00	91	33
01	49	11	43	76	56	67	36	04	02
20	88	35	43	87	46	39	54	51	34

Figure 3.03 Part of a table of random numbers

in groups of three (in Figure 3.03 this would give 845, as circled, 103, 657, 900, 035, etc.)

Other ways of obtaining numbers at random include using telephone numbers, birthdays, rolling a pair of dice and picking numbers out of a hat. The National Lottery uses numbered ping-pong balls for its random numbers!

■ Advantage of random sampling:
   Human bias is completely removed.

■ Disadvantage of random sampling:
   Because choice is haphazard, there may be great gaps in the types of people or places surveyed; some important parts of the total may not be represented in your study.

2 Systematic sampling

This is when points for study are selected at regular and equal intervals for example, every tenth house or shop, every tenth person leaving the car park, every 500 metres along the river, every metre down the beach, etc.

This method is widely used in fieldwork that follows a line – along a road, down a river or across a valley. For example, you may be intending to study: changes in land use from the CBD through the suburbs to the edge of the built-up area; changes in soils and vegetation from the valley floor to surrounding hill tops; how pebble sizes vary along a beach between two headlands. In all these examples you will follow a line or **transect** (see Figure 3.04).

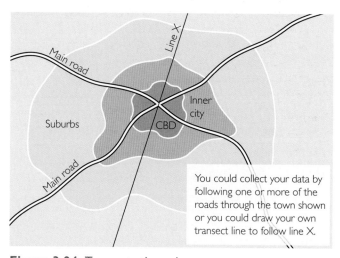

You could collect your data by following one or more of the roads through the town shown or you could draw your own transect line to follow line X.

Figure 3.04 Transects through a town

You could collect your data by following one or more of the roads through the town shown or you could draw your own transect line to follow line X.

Method
You have freedom to choose the distance between sample points depending upon:

- The speed and size of any changes in what you are observing or measuring
- How many sites you think you need to sample
- How many sites you have time to sample.

This is illustrated for a beach transect in Figure 3.05. The student chose the position of the transect down the beach. Every 2 metres, pebble samples were taken. At each sample point, five pebbles were selected using a random sample and quadrats.

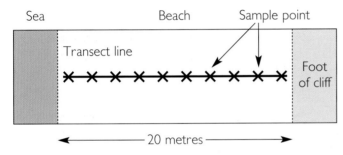

Figure 3.05 A systematic line sample down a beach

- Advantage of systematic sampling:
  There is regular coverage along the full length of study.

- Disadvantage of systematic sampling:
  Not all of the differences/changes may be hit by this automatic selection.

3 Stratified sampling
This is where you pre-select the number of sites or people to be included in your study according to type. For example, if you are investigating a housing area in a town and the census data tells you that it contains 60% terraced, 32% semi-detached and 8% detached houses, you could ensure that your sample observations use similar proportions.

Example – A study of environmental quality of 25 houses. Make sure that 15 are terraced, 8 are semis and only 2 are detached.

Method
Using the example above, let us say the total number of houses in the area is 310. Give each of the 310 houses a number (1–310). Use three figure random numbers to select 25 houses for study. So, using the numbers in Figure 3.03 this would be: 845 (ignore – it doesn't exist), 103, 657 (ignore), 900 (ignore), 035, etc, until the total of 25 is reached. If you find that you have 17 terraced, 6 semis and 2 detached, knock out the last two terraced and go back to the random number tables until the numbers of two more semis come out.

- Advantage of stratified sampling:
  Ensures a fair coverage with random selection.

- Disadvantage of stratified sampling:
  Difficult to set up, making it complicated and time consuming.

TASKS

1 There are 160 separate units on a business park. The student is going to do a 10% sample.

   (a) Use the values in Figure 3.03 to show how the student could randomly select the units to be studied.

   (b) Why is it more likely that, in this case, the student would use a systematic sample?

2 You are going to fill in questionnaires by stopping 50 shoppers as they leave an indoor shopping centre.

   (a) Explain the different ways of choosing the 50 shoppers by (i) random (ii) systematic and (iii) stratified sampling.

   (b) Which would be the best way? Explain.

Preparing questionnaires

Questionnaires are important sources of primary data, especially in human geography enquiries. Using them is the only way to collect information from people. They can be particularly useful for finding out people's opinions, for example, why someone has moved house, what they think about their local shopping area, what their views are about the plan to build a new road, etc. In Figure 3.06 you can find a list of likely topics in human geography that rely greatly for their success upon information from questionnaires.

It is not easy to write a good questionnaire. You need to think carefully about every question and how well other people will understand the questions. Ask yourself: 'Will this question give me answers to help with the aims of my work?' It is always best (it is essential really) to test the questions on friends and relatives first, in order to check that other people are likely to give the type of answers you expected and are going to be able to use.

Read the DOs and DON'Ts list to find out the important points about planning a questionnaire. Look at the example of a generally good questionnaire in Figure 3.07. Pay particular attention to the teacher's comments, which are highlighting its strengths. There are also one or two weaknesses, which proves the point about getting someone to check your questions before printing off large numbers of questionnaires.

QUESTIONNAIRE TOPICS

STUDIES OF THE SPHERE OF INFLUENCE OF A PLACE OR FACILITY

Examples:
- CBD, town or village
- A shopping centre
- A leisure or tourist attraction.

DISCOVERING PEOPLE'S VIEWS AND OPINIONS ABOUT A LOCAL ISSUE

Examples:
- Where should a by-pass road be built?
- Is more car parking needed in the city centre?
- Should planners give permission for building a new supermarket?
- Are there too many tourist visitors?
- Is noise pollution a big problem in the area?

FINDING OUT ABOUT PEOPLE'S HABITS AND WAYS OF LIFE

Examples:
- Where do they shop – why and how often?
- Which places do tourists to the area visit most?
- Journey to work survey.

Figure 3.06 Using questionnaires – likely topics

DOs AND DON'Ts

DO

✔ Keep the questions short

✔ Make sure that every question is needed

✔ Include tick boxes and optional answers

✔ Ask someone to check your questions

✔ Do a pilot questionnaire – try it out first

DON'T

✘ Use direct questions about age or income that will cause embarrassment

✘ Ask for addresses (use the name of the area or postcodes)

✘ Ask questions for which the only answers are 'yes' or 'no'

✘ Include dumb questions like 'Do you come here often?'

✘ Have so many questions that people will become fed up answering them

Hello, I am a student at Hare Hill school. Please could you answer eight quick questions to help me to complete my GCSE geography coursework?

A good start, remember to use it.

1. How often do you visit this shopping area?

Daily ☐ **2/3 times per week** ☐ **weekly** ☐ **once a fortnight** ☐ **less often** ☐

Good opening questions

2. How did you travel here today?

Walk ☐ **Bicycle** ☐ **Car** ☐ **Bus** ☐ **Train/Tube** ☐

3. How long will you stay?

Closed questions with a selection of answers are ideal here.

under 10 minutes ☐ **up to 30 minutes** ☐ **up to 1 hour** ☐ **more than 1 hour** ☐

4. Why do you come to shop here? You can give me more than one reason …

Near to home ☐ **Near to work** ☐ **Choice of shops** ☐ **Pleasant environment** ☐ **Other** ☐

Good idea to have this category.

5. What goods are you buying today?

Food ☐ **Sweets** ☐ **Take away food** ☐ **Stamps** ☐ **Greeting cards** ☐

News-papers ☐ **DIY materials** ☐ **Flowers** ☐ **Shoes** ☐ **Other** ☐

Well chosen. They link in to your shopping area.

6. On a scale of 1–5 (1 is very poor, 5 is excellent) please tell me how you would rate the following:

Well done, you will be able to compare these results with your own survey of shopping quality.

Range and variety of shops	☐1 ☐2 ☐3 ☐4 ☐5
Quality of shops	☐1 ☐2 ☐3 ☐4 ☐5
Parking facilities	☐1 ☐2 ☐3 ☐4 ☐5

7. Where do you normally do your weekly/monthly food shopping?

.... *Maybe better to list some options here.* ....

8. In which street or area do you live?

.... *No need for an exact address.* ....

move thank you up here.

9. Sex (visual)

Male ☐ **Female** ☐

good, not asking but observing.

10. Age (estimated) **0–16 17–25 26–60 over 60**

....

Thank you very much for your help *good.*

A very pleasing questionnaire. Well done.

Figure 3.07 A good questionnaire checked by a teacher

Hi, my names Jo and I want to ask some questions.

Avoid questions about age—or offer a range

1. How old are you? _____ *A little bit rude! Reword it explaining why you are asking the questions.*

2. Where do you live? _____ *England! How precise do you want people to be?*

3. How did you get here? _____

4. Why did you come here? _____

Not everyone understands these terms.

6. Do you buy convenience or comparison goods? _____

7. Where else do you shop? _____ *For what? Food? Clothes?*

8. Do you like shopping here? _____

9. Is it close to where you live? _____ *Work this out from Question 2.*

10. How much money are you going to spend today? _____

11. Do you shop here very often? _____ *Give options, e.g. daily, 2/3 times a week*

12. Which shops have you been to? _____

13. How many marks out of 10 would you give this place? _____ *Too large a range. Is 10 high or low?*

14. Do you think it could be improved? _____ *Avoid yes/no questions. Better to ask 'how'?*

Too many questions

15. How far have you travelled? _____ *Can work out from question 2 again!*

Include a thank you

OK, that's it.

More thought needed Jo, have another go

Figure 3.08 Example of a bad questionnaire

Conducting questionnaires and interviews

Where and when to conduct the questionnaire?

You will generally have a choice as to the location of your questionnaire survey. If you are doing a shopping survey you can either conduct your questionnaire where the shops and shopper are, or you could ask people where they shop in a house-to-house survey. If you are doing a study in a High Street or Central Business District choose your spot carefully; it is best to choose a busy place where people of all ages are passing.

You must also consider the timing – which days of the week and at what times of day? Midweek shoppers may have very different profiles from the Saturday shoppers. Indeed this could form the basis of an interesting study in itself: 'How do the characteristics of shoppers vary on different days and at different times in the week?' If you are doing questionnaires in two or more places, try to do them on the same day or at similar times. You can then compare your results more easily.

How many questionnaires do you need?

This is like asking 'How long is a piece of string?' It depends upon how many questions you are asking, how detailed they are and whether the questionnaire is the main method of data collection. If you are undertaking a relatively simple shopping survey, you could be expected to do between 50 and 100. With a more detailed questionnaire 20 to 30 may be sufficient. If the questionnaire is a very central part of your work and the most important form of data collection, then you would be expected to have collected a significant number – certainly 100 questionnaires would be considered a significant number. If the questionnaire is just one of several techniques then fewer are acceptable.

TOP TIPS

The number of questionnaires

- The more the better (provided time allows).
- If you use 25 or 50 or 100, it is easier to calculate percentages from the results.

Interviews

These are more detailed than questionnaires. In most coursework enquiries there are likely to be no more than five interviews, ten at the most. Often there are only one or two when they are being used to supplement other methods of data collection. You need to prepare your questions in the same way as for the questionnaires, because the person being interviewed has only a limited amount of time to spend with you. Your aim should be to obtain the largest amount of information in the shortest possible time. It may be a good idea to tape the interview so that you can write down the answers later, but do make sure you ask the interviewee's permission first each time.

TOP TIPS

Good technique for conducting a questionnaire

- Dress smartly and be polite.
- Explain why you are doing the survey.
- Record the place, date and time.
- Don't block the path or entrance.

TASKS

Main task – to explain why the questionnaire in Figure 3.08 is a bad one to use for a survey of shoppers.

1 Explain what is wrong with the introduction and the ending.

2 Which questions are likely to lead to yes or no answers?

3 Which questions may appear rude and why?

4 Give an example of a question that most shoppers will not understand.

5 (a) What is wrong with question 13?

(b) How might the question be improved?

6 (a) Why will question 15 not be answered accurately by shoppers?

(b) Why should the geography student be able to work this answer out more accurately than the shopper?

7 Are there too many questions? Give your views on this.

Environmental quality surveys

Whether or not you are studying an urban or a rural environment you can assess the quality of the landscape by using an environmental quality survey. This involves giving a score (or mark) to a whole series of landscape features. The larger the score at the end of the exercise, the more impressive or the higher is the quality of the landscape. The results are your own perception of the landscape. You may take someone else along to do the survey too. This will provide an alternative viewpoint. You could then take the average score for the landscape.

Figure 3.09 is an example of an environmental quality survey in an urban area.

	Positive evaluation	2	1	0	−1	−2	Negative evaluation
1 Housing layout and design	Varied and interesting, well spaced						Poor unimaginative, high density
2 Building materials	Attractive						Drab and uninteresting
3 Natural features	Trees and grass improve appearance						No trees and grass
4 Open space	Present and providing safe play areas for children						Absent with nowhere for children to play
5 Gardens	Present and well maintained						No gardens or poorly maintained
6 Car parking	Parking mainly off the roads						Parking mainly on roads
7 Pavements	Present, adequate and safe						Not present or inadequate
8 Kerbs	Low and easy to use with pushchairs						High and inconvenient
9 Traffic noise and fumes	Quiet, not a problem						Noisy, a major inconvenience
10 Road crossing	Little traffic, slow moving, easy and safe to cross						Busy roads, difficult and dangerous to cross
11 Litter	No obvious litter						Heavily littered
12 Vandalism	No obvious vandalism						Property damaged and abused, graffiti

Figure 3.09 An example environmental quality survey for an urban area

Look at the categories closely – you may need to remove some factors and add in other qualities that are more appropriate for your title and aims. If you are studying the impact of a new development, you could fill in two copies of the sheets – one for before the development took place and one for after. This technique for study before and after a change can also be used for questionnaires. Do another survey using a similar questionnaire after the change has occurred.

Method

- Visit the different areas you wish to assess.
- Make sure you have enough copies of your survey sheets.
- Assess each area and record your results.
- Add up the total score for each environment – the higher the score, the better the quality.

Figure 3.11 The old jewellery quarter

Figure 3.12 Modern business park

TOP TIP

This technique is best used when two or more landscapes need to be compared.

Street Quality Survey

Litter	No litter	3
	Small amount of litter	2
	Much litter	1
	All kinds of litter scattered widely	0
Care of roads and pavements	Well maintained	3
	Slightly uneven	2
	Uneven	1
	Very poor condition	0
Trees, shrubs, grass verges	Well kept	3
	Badly kept or poor quality	2
	Damaged trees and shrubs, overgrown verges	1
	Derelict and unplanted areas	0
Street furniture (lamp posts, seats, telephone boxes, etc)	Essential items well designed and in good repair	3
	Adequate but badly designed seating/litter bins, etc.	2
	Essential items missing or inadequate	1
	Badly cared for or vandalised street furniture	0
Traffic	Clear roads with light traffic	3
	Traffic flowing freely; light parking	2
	Traffic flowing freely; heavy parking	1
	Traffic congested – not flowing freely	0
Noise	Low level noise	3
	Slight traffic and other noise, not too disturbing	2
	Frequent disturbing and distracting noises	1
	Continuous disturbing and distracting loud noise	0
Road signs	Well placed and visible	3
	Badly placed	2
	Confusing and cluttered	1
	Inadequate information	0

Figure 3.10 An example environmental quality survey for a street

TASKS

Do a trial environmental quality survey of the industrial landscapes in Figures 3.11 and 3.12. These show two industrial landscapes in inner-city Birmingham.

1. Think of five headings for an environmental quality assessment. Think about headings that are suitable for industrial areas.

2. Write down what you are looking for as either high and low scores or positive and negative evaluations.

3. Do a trial assessment based on what you can see on the photographs.

4. Compare your results with those of a friend.

5. Decide the assessments that should be used to set the standard.

Photographs

It is not always easy to record in a table or put into words what your eyes can see when you are visiting a place and doing fieldwork. This is where photographs and field sketches (pages 32–33) can be helpful. Photographs are also useful in adding colour and interest to your finished work. They help the people who are marking and moderating your work to visualise what your study area was like. You can also use photographs when writing up the *Data collection* part of the coursework to illustrate how measurements and surveys were undertaken. They are a very useful record of what you have seen. Look at Figure 3.13, which shows how to make good use of a photograph.

It is now easier than ever to take good quality pictures because of digital cameras. After taking every picture you can check whether it shows what you wanted to capture and that the picture is clear and well focused. If it isn't good enough, you simply take another picture. Digital cameras often take better pictures in poor light than ordinary cameras do, an important advantage in countries like the UK where cloud and rain are frequent visitors!

Digital photography has another big advantage for coursework. Since it is downloaded into the computer, it is easy to insert photos anywhere within your written report. An even bigger advantage is that you can highlight certain significant geographical features. Blurred photographs, with key features missing, should have become a thing of the past in completed coursework reports.

TOP TIPS

- Label your photographs.
- Make sure the labels are related to your topic of study.

Ice rounded landscape

Steep sides with bracken, heather, deciduous trees

Mainland – Kintyre

Linear settlement of Lochranza

Fjord-drowned valley

Ruined Lochranza castle

Small, 9-hole golf course

Flat floor of glaciated valley

Bracken in March

Gorse

Rough grazing land

Figure 3.13 Photograph with annotations to highlight the main features of the landscape

Uses of photographs

1 **They help to set the scene.**
They show some of the features of your study area, which can be used in your *Introduction*.

2 **They illustrate how you collected your data.**
They can be used to show you taking measurements or handing out questionnaires, which can support the *Data collection* part of your writing up.

3 **They give examples to show what you mean.**
They can show pebble sizes on a beach survey, or examples of high and low scoring landscapes in an environmental quality survey. This is also useful under *Data collection*.

4 **They provide views of physical features that are being studied.**

Examples include river channels, valleys, cliffs, beaches, natural tourist attractions such as waterfalls, vegetation cover and soil profiles. These are most likely to be included within your *Analysis*.

5 **They show what different human features look like.**
Examples include types of shops and houses, factories and land uses in the country areas. These will also most likely be included in the *Analysis*.

In Figure 3.14 the labels in black are for setting the scene; whatever the theme of your study, you could use these labels. The labels in blue would be useful to a physical geography study. The labels in red are the ones that could be useful to a human geography study about tourism.

Figure 3.14 Labelled photograph of Malham in the Yorkshire Dales National Park

Field sketches

These are sketches of landforms or landscapes made in the field. They record your observations. To be useful they should be well drawn, but at the same time keep them simple. They must be labelled in order to identify the main features of a landscape. Field sketching is an important geographical skill, highly rated by coursework moderators. Use colour on your final neat version to improve the appearance. Always refer to your sketches when writing up the work.

Basic guidelines for drawing field sketches

- Draw a box as a starting point for your field sketch.
- Draw in the skyline first, using a pencil.
- Mark on the major features to provide the fixed points – sometimes it is easier to make a frame with your fingers and hold it at arm's length (Figure 3.15).
- Sketch in any minor features that are relevant.
- Label your sketch in the field – don't wait until you are back at home. It is amazing how much you will forget.
- Add a title, record the location and direction of view.

An example of a physical field sketch, and a photo of the same location, are given in Figures 3.17 and 3.16. Notice

Figure 3.15 Frame your sketch

how the field sketch simplifies the scene and allows you to concentrate upon the features you are interested in.

Field sketches have the same uses as photographs. A field sketch has the advantage over a photograph in that it is easier for you to select what is significant from everything you are observing. Your labels give it extra impact.

Look at Figure 3.18, which is a photograph of an inner city area. Figure 3.19 is a student's sketch from the same viewpoint. The student's topic of study was 'change in the inner city'. Notice how the student, by concentrating upon only a few houses, has been able to pick out and label those features of the houses that show change (in this case improvements).

Figure 3.16 Photograph of the Dorset coast at Lulworth

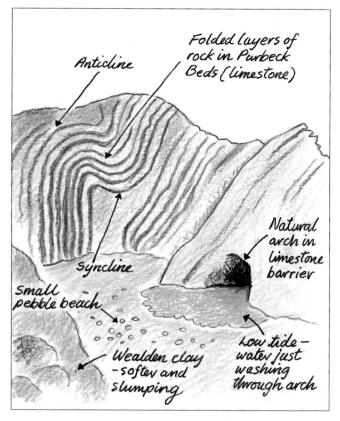

Figure 3.17 Field sketch of the Dorset coast at Lulworth

Figure 3.18 Housing in an inner city area

hanging flower basket new windows fresh paint

Figure 3.19 Field sketch to show changes

TASKS

Practise drawing a sketch.

1 Draw a sketch of Figure 3.20.

2 Add labels for a study with the title 'Landscape features attractive to visitors'.

Figure 3.20 Horton in Ribblesdale and Pen-y-ghent in the Yorkshire Dales National Park

Checklist

Make sure that your field sketch has the following:

○ *Title*

○ *Details of location*

○ *Labels for features important for study*

○ *Date drawn* (if significant)

Other methods of data collection

So far in this chapter the emphasis has been upon primary data collection using questions and observation. The eagle-eyed ones among you will have spotted that there has been nothing so far on measurement. The explanation is that techniques of measurement are more topic-specific, especially those which need special equipment. They are much more widely used in studies of physical geography. This is why you will find them in Chapter 6 among the relevant topic examples. A summary of measurement techniques that are referred to later in the book is given in the box below.

For some of these there are both hi-tech and low-tech ways of taking the measurements. For example, to measure the speed of river flow, a flow meter with digital controls will give an instant accurate reading. Using floats such as biscuits and kiwi fruit, and timing how long it takes them to travel over a measured distance, not only takes longer but needs to be done three or four times to ensure even reasonable accuracy. However, which method is more fun?

Coursework with an environmental flavour is something of a half-way house (for environmental pollution see pages 100–101). To measure air pollution accurately and identify the pollutants, sophisticated electronic monitoring instruments are needed, which schools do not possess. Likewise, to discover what is causing water pollution needs careful laboratory investigation. However, visible signs of water pollution and differences in intensity between places can be estimated by observation; similarly types and amount of litter can be observed. These are more likely avenues for coursework enquiry.

Figure 3.21 Much fieldwork in physical geography relies upon the availability of equipment, although it doesn't need to be at the cutting edge of technology. Taking precautions to keep warm and dry is also important.

In topics in human geography, counts are more likely than measurements. The two most commonly used in urban studies are traffic and pedestrian counts (see page 84). Sometimes techniques more associated with physical geography are used. Look at Figure 3.22 which shows an example of quadrats being laid across the width of the path and being extended on to the sides of the path as well; their use will allow students to estimate the percentage of ground within the quadrat that is vegetation covered and to work out the width of footpath erosion at this point. Repeating the process further along the path allows the extent and severity of the erosion to be measured.

MEASUREMENT TECHNIQUES

River studies (pages 66–69)
- Width and depth of channel
- Speed of flow
- Discharge
- Gradient
- Load size and shape

Coastal studies (pages 74–77)
- Height of cliffs (and other landforms)
- Angle of slope of beaches
- Wave frequency and height
- Size of beach materials

Weather (pages 94–95)
Measurements at weather stations for:
- Air pressure
- Temperature
- Precipitation
- Humidity
- Wind speed

Vegetation and soils (pages 92–93)
- Using quadrats for percentage amount and types of vegetation cover
- Depth of soil and width of soil horizons
- Testing for colour, texture and pH of soils

Figure 3.22 Using quadrats to measure footpath erosion

Using secondary data from the Internet

The Internet is at the cutting edge of technology. Despite this, access to it is readily available in homes, schools and colleges. No longer is it essential for geography students to go on a physical search of libraries and offices in order to borrow books and look at reports; now they can be brought to you on-screen. Search engines save you hours of effort. They can also locate data that is more up to date than anything that is printed and published.

However, there is a downside. The temptation to browse too much and download impossible amounts of information is ever present. Look at and obey the DOs and DON'Ts about using the Internet below.

Use of secondary data is always the most effective when it supports the primary data that has already been collected on the same topic.

Examples

1   Data collected by fieldwork about housing types in two areas of the city.

Possible Internet sources:

A The 2001 Census – for data about housing in the whole of the city, so that you can discover similarities and differences between your areas and the rest.

B Estate Agents' websites – for data about housing prices in your areas, as well as in other areas of the city.

2   Variations in visitor numbers in a coastal resort/National Park.

Possible Internet sources:

A Weather station data (both past and present) from the nearest station.

B Newspaper websites for weather charts and reports on fieldwork days.

C Tourist Office and National Park Authority visitor numbers.

3   Coastal studies in locations where erosion is a local issue.

Possible Internet sources:

A Local council for details of coastal protection works and coastal management.

B Websites of pressure groups formed by local people concerned about the effects on them of more coastal erosion.

DOs AND DON'Ts

DO	DON'T
✔ Select information and use only that which is relevant to your coursework	✗ Use words or information that you do not understand
✔ Remember that the worked is marked for quality not quantity	✗ Include all the information and data that you download
✔ Do name the source of your information	✗ Overload your report with masses of detail

Chapter 4
Presenting data

Introducing the techniques

Figures 4.01 to 4.04 illustrate the most commonly used methods of presenting and processing in geography coursework. Why are maps, tables and graphs much loved and greatly used by geographers?

- Maps are vital for showing the location of the study; they also show patterns on the earth's surface and changes from one area to another.

- When a lot of values are placed in a table, the data is easier to see and to understand.

- Graphs summarise data in ways that make it easier for you to see what is important. It is easier for you to pick the main trends and any unusual values that do not fit the general trend.

Methods for presenting and processing data in geography coursework

Maps
No geography coursework should be without them. Maps show locations and patterns. See pages 42–43.

Tables
Also included are lists and simple statistics. If you have taken measurements, or used questionnaires, or have done environmental quality surveys, tables will be needed. See pages 44–45.

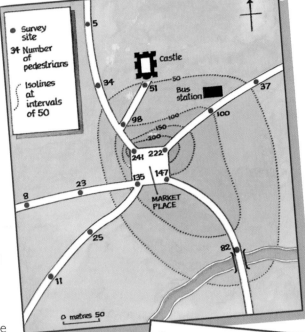

Figure 4.01 Map of isolines

1 In which place do you work?		Total
Durham	ЖЖ ЖЖ ЖЖ ЖЖ	20
Newcastle-upon-Tyne	ЖЖ	5
Sunderland	IIII	4
Washington	III	3
Peterlee	II	2
2 By which means of transport do you travel to work?		
Car	ЖЖ ЖЖ ЖЖ ЖЖ ЖЖ	25
Bus	ЖЖ I	6
Walk		0
Other	III	3
Note that the answers to question 1 were used to plot the star diagram on page 40.		

Figure 4.02 Tally chart for a journey to work study

Tackling Geography Coursework

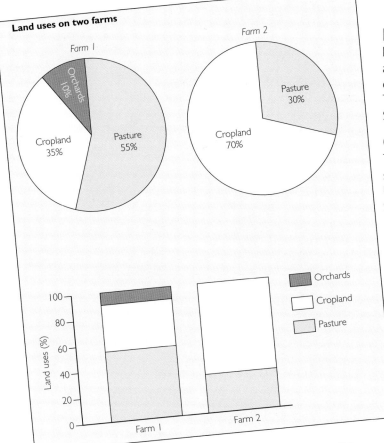

Figure 4.03 Pie and divided bar graphs showing rural land uses

Photographs and field sketches

Neither are absolutely essential, but if annotated and well placed, they help to bring your coursework to life.

They can make a big difference.

See previous pages 30–33.

Graphs and related diagrams

These are really versatile, they can be used to show many different types of data.

Use them.

See pages 38–41.

Statistical work

This is dreaded by many.

Don't worry, it is optional.

A few love it, however, and will have collected the type of data which lends itself to further statistical treatment. See pages 46–49.

From the many different ways to present and process data, it is your job to pick those that are appropriate to your needs. Ideally you will have collected sufficient data, with enough variety, to allow and encourage you to use a variety of methods. Within the statistical section, although it may be of little interest to a lot of you, there are descriptions for two other types of graph that might be useful to some of you. They are the dispersion diagram (page 46) and scatter graph (Figure 4.04, right).

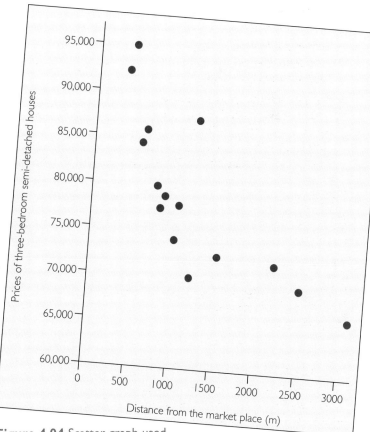

Figure 4.04 Scatter graph used in a study of housing

Most widely used graphs: line, bar, pie and pictograph

Very few pieces of coursework are completed without the use of one or more of these. In fact, it is not uncommon for all four to be included because they have so many different uses. Line graphs are useful for showing changes with time. Bar graphs show totals. Divided bar graphs and pie graphs display relative sizes of the different parts that make up the total. Pictographs show the same but in a more visual manner.

Line graph

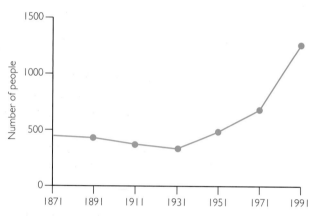

Figure 4.05 Example of a line graph – village population

How do you draw it?

■ Plot the values on the axis up the sides (the *y* axis).

■ Look at the size of the values to be plotted.

■ Choose a scale which will fit in the space and will show differences between the values.

■ Draw two axes; the time axis (with dates, etc.) goes along the bottom (the *x* axis).

■ Label your axes.

■ Plot your values with a clear dot or cross where the values meet on the two axes.

■ Join up the dots or crosses with a line.

When can it be used?

■ To show climate and weather features such as temperature (but not rainfall).

■ For population data such as changes in numbers with time.

■ Production data such as car output or crop yield, i.e. when you have data for different dates or for different times of the year.

Bar graph

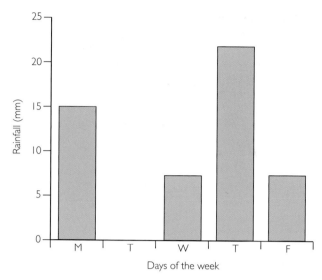

Figure 4.06 Example of a bar graph – rainfall record at a school weather station

How do you draw it?

For a vertical bar graph as shown above:

■ Check the size of your values.

■ Draw a frame with the two axes.

■ Label your axes.

■ Plot a scale which covers all the values on the vertical axis (the *y* axis).

■ Draw bars of equal width up from the base on the *x* axis so that the top of the bar reaches its value on the scale.

When can it be used?

The type shown above is useful for showing any kind of total value. A vertical bar graph is the only way of showing rainfall.

A divided bar graph is different and is used to show the parts that make up the total – like a pie graph. The bars can also be drawn so that they are horizontal; this makes labels easier to read. In population pyramids, horizontal bars are always used.

Figure 4.07 Example of a divided bar graph – land uses on two farms

Pie graph

Land uses on two farms

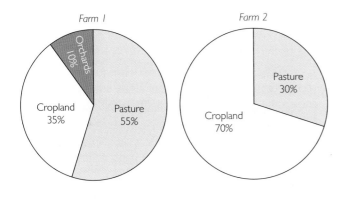

Figure 4.08 Example of a pie graph – land uses on two farms

How do you draw it?

■ Change the values into percentages.

■ Multiply each percentage by 3.6°:
   e.g. 1% = 3.6°   10% = 36°

■ Draw a circle of convenient size.

■ With a protractor, plot the largest segment first, starting at 12 o'clock.

■ Go in a clockwise direction from large to small segments.

■ If there is a sector for 'others', plot this last.

■ If you are drawing more than one circle, keep the sectors in the same order irrespective of size.

■ Colour or shade in the segments and give a key for them.

When can it be used?

Whenever there are parts for the total. Examples include:

■ Housing – percentages of different types such as terraced, semis and detached

■ Traffic counts – percentages of cars, lorries, buses and bikes

■ Land use – percentages of woodland, grassland and cropland

■ Vegetation survey – percentage of the ground covered by different types of plants

■ Beach survey – percentage of sand and of pebbles of different sizes

■ Shopping centre study – percentages of shoppers who are local.

Pictograph

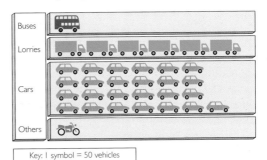

Key: 1 symbol = 50 vehicles

Figure 4.09 Example of a pictograph – results of a traffic survey

How do you draw it?

■ Choose a symbol that looks like what you are trying to show.

■ Look at the size of the values to be plotted.

■ Work out the number to be shown by one symbol, for example, one symbol for every 50 counted in the traffic survey.

■ Each symbol used needs to be drawn the same size (except if a part symbol at the end is essential).

When can it be used?

■ For plotting any numbers observed or counted such as people, pebbles, parking spaces, etc.

■ For comparing single sets of figures such as percentages.

These are the most widely used graphs because they can be used to show such a variety of data collected, either from fieldwork or from other secondary sources such as the census or weather station records.

TASKS

Climate data for York	J	F	M	A	M	J	J	A	S	O	N	D
Temperature (°C)	4	4	6	9	12	15	17	17	14	11	7	5
Precipitation (mm)	59	46	37	41	50	50	62	68	55	56	65	50
Number of wet days	17	15	13	13	13	14	15	14	14	15	17	17
Percentage of wet days in a year – 48.5%												

1 (a) Name the only techniques that are used for showing (i) temperature (ii) precipitation.

   (b) Can you explain (i) why the two techniques are different and (ii) why no other techniques are considered to be appropriate?

2 (a) Name a technique suitable for showing the number of wet days per month.

   (b) Explain why it is suitable.

3 Draw a graph to show the percentage of wet days in a year at York.

Other ways of presenting data

Block bar graphs and proportional circles are more complex techniques compared with the simple bar and pie graphs upon which they are based. However, there are many different types of data for which they are suitable methods of presentation. They should be considered when you are looking to add variety to your presentation, and are methods that might be considered to be more complex by the moderator/examiner of your work.

Block bar graph

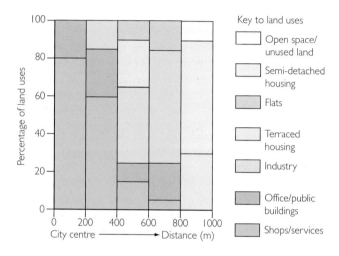

Figure 4.10 Example of a block bar graph – changes in land use in a UK city

How do you draw it?
- Draw two axes.
- Distances are marked along the horizontal axis.
- Percentages are shown on the vertical axis.
- Make a block of separate vertical bars of the same height and width.
- Draw a key for colours/shading for the different parts which make up the total.
- Plot the percentages for the different parts in each bar.

When can it be used?
For continuous land use surveys along a route or transect, for example, from the centre to the edge of a town. In the example in Figure 4.10, land uses were observed and summarised every 200 metres as percentages. It can also be used for land use surveys in rural areas, for example, across a river valley or from the valley floor to the top of the hill.

Proportional circles

How do you draw it?
- Find the square roots of the total values to be shown. (The radius of the circle must be in proportion to the square root of the value to be plotted and not to the size of the value itself.)
- In deciding upon the unit size for each radius, remember that it is not easy to draw small circles accurately, nor to make segments within them for a pie graph.
- Don't make the unit size for each radius too large so that the circles do not fit on the page or the map.
- Draw your circles with a compass once you have decided upon a satisfactory scale.

When can they be used?
For plotting totals of any kind.

In addition, if the parts that make up the total are known, the inside of the circles can be used for a pie graph.

Usually the quickest and easiest method of showing total values is the bar graph, but if you wish to demonstrate that you are capable of using a more complex method, try the proportional circle for a change. If you have limited space for drawing the graphs, such as on a map, the circle is likely to be the better method.

Star diagram

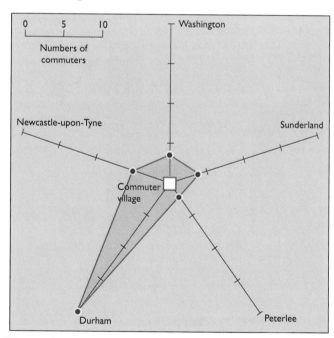

Figure 4.11 Example of a star diagram – where commuters from the village work

How do you draw it?

- Calculate the number of degrees to give equal spacing between the different classes or types to be shown.

- In the example in Figure 4.11, there were five places of work to be shown; 360° divided by 5 = 72° difference for the spacing.

- Choose the central point for your diagram and mark it clearly.

- Draw lines out from your central point, one for each type being shown.

- Look at the size of the quantities to be plotted and think about a suitable linear scale.

- Make each line proportional in length to the quantity it represents.

- Join up the ends of the lines by straight lines (which will give you something like a star shape, as the title suggests).

When can it be used?

- For showing what people do or what people think about issues and problems.

- For plotting many different kinds of economic data.

- Note that Figure 4.11 was drawn to show the same data as was used in the tally chart (Figure 4.19 on page 44) and on the flow line map (Figure 4.15 on page 42).

Kite diagram

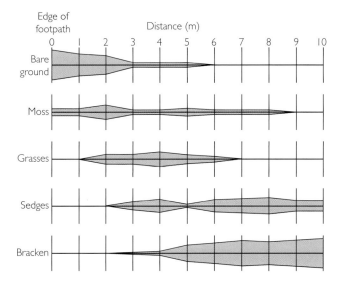

Figure 4.12 Example of a kite diagram – vegetation cover from the edge of a public footpath in a National Park

How do you draw it?

- Make a scale line for the distance covered in the survey (10 metres in Figure 4.12).

- You need one row for each type of plant/vegetation cover (there are five of these in Figure 4.12).

- Each row needs to have the same scale and be wide enough to allow 100% for each plant/vegetation cover.

When is it used?

Kite diagrams are used most often in studies of vegetation, when making observations of plants at regular points along a transect line. This can be for part of a land use study. The example used here was drawn for an enquiry into the impact of visitors in a National Park.

TASKS

1 State the percentage values shown in Figure 4.13.

2 Draw a graph of the same type to show land uses 800–1000 metres from the city centre from Figure 4.13.

3 **(a)** Name the two other types of graph for which this is an alternative method of presentation.

   (b) Why is the use of alternative techniques from time to time in coursework reports helpful?

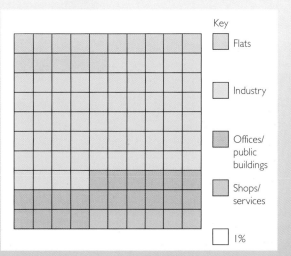

Figure 4.13 Land uses 600–800 metres from city centre

Maps for showing data

Choropleth (shading) map

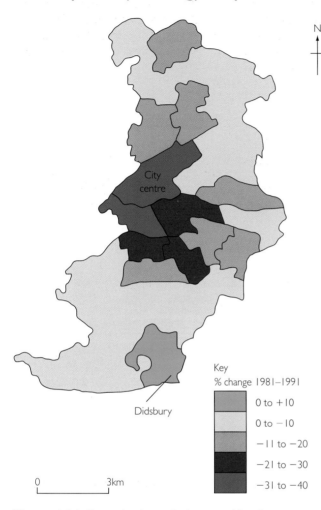

Figure 4.14 Example choropleth map – Manchester population change 1981–1991

How do you draw this type of map?

- Note down the highest and lowest values.
- Split them up into 4 or 5 classes (groups) of equal size.
- Use 0 (zero) as a class boundary if there are both negative and positive values.
- Choose a colour or type of shading for each class. Choose darker and denser colours for the higher values.
- Use a different colour for the negative values.

When can it be used?

- To map percentages for areas.
- To map values per unit of area such as people per sq km or the number of houses per hectare.
- Secondary sources of data are often used (in Figure 4.14 the source was the census).

Flow line map

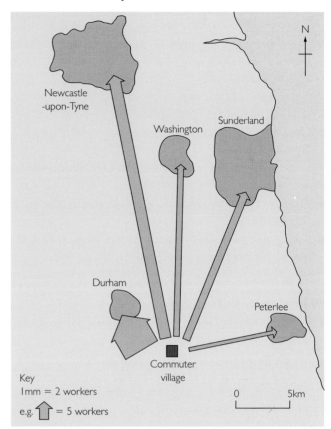

Figure 4.15 Example flow line map – where people from the village work

How do you draw this type of map?

- Look at the size of the values and the space on your map.
- Decide on a scale for the width of the lines that will fit (in Figure 4.15 this was 1 mm for two workers).
- Plot lines of varying width between village and places of work.
- Line widths vary according to the number of workers.

When can it be used?

- To show any kind of movement between two or more places, for example, traffic flows, pedestrian flows, number of shoppers, etc.
- The inside of the lines can be sub-divided, for example, the flow lines on Figure 4.15 could show male and female, or different means of transport used by commuters.

Desire line maps

Desire line maps and star diagrams are similar, but desire lines are placed on a map rather than a diagram.

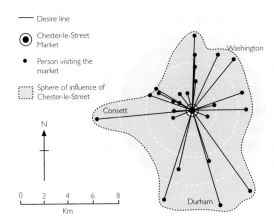

Figure 4.16 Example of desire lines – people visiting Chester-le-Street market

How do you draw it?

- On the map, highlight clearly the place where you did the survey (place of work, shopping centre, sports centre, tourist attraction, etc.).

- Mark by a dot on the map the place from which the first person in the survey came.

- Draw a single straight line between the survey point and the place the person came from.

- Repeat this for the journeys made by all the people in the survey.

- When you have finished, you can draw a line around the outside edge of the desire lines to indicate the limits of the sphere of influence of the place or service being surveyed.

When can it be used?

- For showing information about journeys made, whether by people or goods or by different means of transport.

- For showing the sphere of influence of one or more of the following – settlement, city centre, markets, shop, shopping centre, factory, places of recreation such as parks and sports centres, tourist attractions, bus and railway stations.

- This type of map is likely to be useful when you have collected data by questionnaire and asked questions at your survey point such as 'Where do you live?' or 'Where have you travelled from?'

Isoline map

An isoline map is the most difficult type of map to draw. All the rules for drawing it must be obeyed, otherwise the map is no good. Quite a lot of values are needed in order to be able to detect a pattern from which the isolines can be drawn. It is regarded as a highly complex technique. A contour map, with lines linking places of equal height, is the best-known type of isoline map.

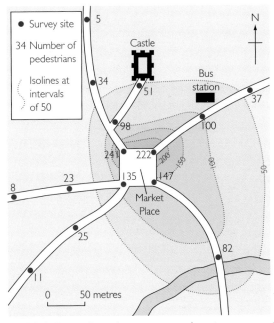

Figure 4.17 Example isoline map – pedestrian counts in the town centre

How do you draw it?

- On your base map, write down the numbers counted at each survey point.

- Decide upon a regular interval for the lines. (In Figure 4.17 the isoline interval is 50 people).

- Draw in the lines (as on Figure 4.17) bearing in mind that:

  - Lines go through survey points where the number counted was the same as the number of the line

  - All points of lower value than the line value must lie on one side of the line and all higher values on the other

  - All lines must be shown even if there is no direct evidence for plotting them.

- Write the values on the lines.

When can it be used?

- For traffic and pedestrian counts.

- For plotting temperatures in micro-climatology studies.

- For average journey times to reach work, town centres, football grounds, etc.

Tables and other ways of summarising data

Successful data collection typically gives you lots of figures. Indeed, if your data collection has not generated a mass of figures, it might suggest that you have not collected enough.

Examples

- If your study was physical, in a river or along the coast, you will have many measurements.

- If you handed out questionnaires, you will have counted up total numbers.

- If you used a source of secondary data, such as the census, you will have values galore to process.

SHOP/SERVICE	NUMBER
1 Eating places	12
Pubs and clubs	3
2 Building societies	7
Banks	5
Travel agents	4
Post office	1
3 Department stores	5
Stationery and books	4
Clothes shops	7
Shoe shops	8
Bakers	3
Music shops	3
Other shops	13
TOTAL	75
CALCULATIONS	

1 % of eating and drinking places

$$\frac{15}{75} \times 100 = 20\%$$

2 % of financial places

$$\frac{12}{75} \times 100 = 16\%$$

3 % shops

$$\frac{43}{75} \times 100 = 57.3\%$$

Figure 4.18 Table to show the number of shops and services in a town centre

The record sheets you filled in whilst doing your fieldwork should not be included in the main part of your coursework. At most, one example of a completed data collection sheet or questionnaire should be included. Therefore one of your first tasks is to use the recording sheets' raw data to make a neat version of all the data collected. Making tables is one of the easiest ways of doing this.

Tables

These can be of any size or shape, provided that within it the data are laid out in a clear way. Clarity is all-important. Don't be afraid to highlight within the table the most important data for your study. Very few students do this. Within a table only brief headings are possible; add labels where there is room around the edges to point out the exact nature of the data. Don't overload your tables; they are supposed to be there to make things clearer, not to confuse. Use more than one table if necessary.

Once you have your values in the table, it may be possible to do some simple calculations such as:

- Adding up the totals
- Working out some average values
- Converting some of the values into percentages.

Tally charts

These are useful as a first stage in sorting data. They are most useful as a means of sorting and summarising data counts from questionnaires (see Figure 4.19).

1 In which place do you work?		Total
Durham	ЖЖ ЖЖ ЖЖ ЖЖ	20
Newcastle-upon-Tyne	ЖЖ	5
Sunderland	IIII	4
Washington	III	3
Peterlee	II	2
2 By which means of transport do you travel to work?		
Car	ЖЖ ЖЖ ЖЖ ЖЖ ЖЖ	25
Bus	ЖЖ I	6
Walk		0
Other	III	3

Figure 4.19 Tally chart of results from a journey to work questionnaire

Note that these questionnaire answers were used to plot the map that is Figure 4.15.

Data summary – an example

Look at Figure 4.20.

Column 1 contains raw data – the prices of houses within a city are listed in the order they were taken from property advertisements in the local newspaper. As they aren't in any order, they don't tell you very much.

In Column 2 the house prices have been ranked, i.e. they have been put in order of price from highest to lowest. Some organisation of data has occurred.

In Column 3 some simple calculations have been done. Now it is easier to see some of the significant points from the data.

If your data collection includes numbers, you should do simple calculations such as those shown in Column 3 of Figure 4.20. Always remember, however, that these are just statistics. They are a great geographical aid, but are not in themselves geographical. Statistical results such as these suggest further study opportunities of a geographical kind.

For example, you may wish to concentrate your investigation upon those houses above the average price. Where are they located? What geographical reasons help to explain why they are so highly priced? Or you may wish to study the distribution of one type of house and do an environmental quality survey to help to discover geographical reasons for price variations for that type of house.

TASKS

1 (a) Name the types of graphs, diagrams or maps that could be used to show the data in Figures 4.19 and 4.20.

(b) Explain why the methods named are appropriate.

2 (a) What does it mean when data is 'ranked'?

(b) Refer to Figure 4.20. State and explain two advantages of ranking the house prices.

1 Raw data: price of houses Source: local newspaper			**2**				**3**
House type	No. of bedrooms	Price (£)	House type	No. of bedrooms	Price (£)	Ranked prices of houses	Simple statistical values calculated from the data
TH	4/5	410 000	T	2	111 900	1. 410 000 16. 190 000	The range of price is highest to lowest = £410 000 to £111 900
D	4	350 000	T	2	123 000	2. 398 000 17. 185 000	The mean (average) price is: total of the prices divided by number of houses = £225 590
D	4	320 000	TH	4/5	159 950	3. 359 900 18. 319 900	The median (middle) price is middle of the ranked values = £194 000
TH	3	171 900	SD	3	140 000	4. 350 000 19. 169 000	
TH	3	157 900	D	4	299 000	5. 350 000 20. 159 900	Mean prices for different types of houses are:
TH	4/5	279 000	T	2	113 900	6. 320 000 21. 157 900	4/5 bed town houses = £249 770
TH	4/	119 000	SD	3	159 900	7. 319 900 22. 145 000	4 bed detached = £337 000
T	2	118 000	D	4	398 000	8. 319 900 23. 140 000	3 bed town houses = £171 600
SD	3	145 000	SD	3	139 000	9. 318 000 24. 139 000	3 bed semi-detached = £153 700
SD	3	133 000	SD	3	190 000	10. 299 000 25. 137 800	2 bed terraced = £121 280
D	4	319 000	TH	4/5	359 900	11. 279 000 26. 133 000	
TH	4/5	319 900	TH	4/5	350 000	12. 230 000 27. 122 600	
TH	4/5	210 000	TH	4/5	230 000	13. 210 000 28. 119 800	
SD	3	169 000	TH	4/5	199 900	14. 199 900 29. 118 000	
TH	3	185 000	T	2	137 800	15. 198 000 30. 111 900	

TH Town house T Terraced D Detached SD Semi-detached

Figure 4.20 House prices in London

Using statistics to process data

You were introduced to some simple statistical processing of data on the previous page. If your data collection has provided you with a lot of values, you might gain from going a stage further than this with statistical processing of data.

The first two stages in data processing are:

1 Find an average value.
2 Examine the spread of values around this average value.

Figure 4.21 contains raw data listing prices for the same type of house in two areas within the same town. This data will be used to illustrate the calculations for averages (median and mean) and for the spread of values (quartiles and standard deviation).

Area A
72 500: 66 500: 84 500: 70 000: 79 950: 95 000: 85 950:
78 950: 92 500: 69 500: 87 500: 71 900: 73 950: 75 900:
77 500

Area B
54 950: 53 950: 51 950: 53 500: 50 000: 55 950: 49 950:
52 500: 53 950: 57 950: 48 500: 55 000: 53 250: 54 750:
52 950

Figure 4.21 Prices (£) of three-bedroom terraced houses for two areas in the same town. Source: Estate Agents Advertisements

Median, quartiles and index of variability

The quickest method, involving only a few calculations, is to use the median in stage 1 (the average) and to calculate quartiles, inter-quartile range and index of variability in stage 2 (spread of values around the average).

Working out the median

The **median** is the middle value. There is a total of 15 values for each housing area; therefore the median is the eighth value working either from the bottom or the top. The median can be obtained by ranking the values (from highest to lowest) or by using Figure 4.22.

Median for area A = £77 500
Median for area B = £53 350

HOW TO DRAW A DISPERSION DIAGRAM

- Make two columns (for the two areas A and B).
- Choose a scale which covers the range of values. (With this method you don't have to start at 0 if it is not needed.)
- Plot each house price in the columns with a dot.

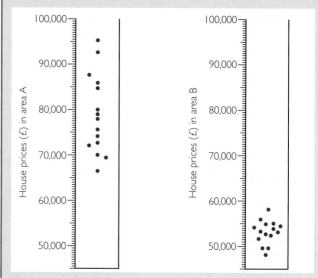

Figure 4.22 Dispersion diagram for the house prices in Figure 4.21

Working out the quartiles

The **upper quartile** is the value with 25% of the values above it and 75% below it.
Therefore the upper quartile for area A is £85 950; area B is £54 950

The **lower quartile** is the value with 25% of the values below it and 75% above it.
Therefore the lower quartile for area A is £71 900; area B is £52 500

The **inter-quartile range** is obtained from upper quartile minus lower quartile.

Therefore

the inter-quartile range for A is 85 950 − 71 900 = 14 050
the inter-quartile range for B is 54 950 − 52 500 = 2450

The inter-quartile range covers the middle 50% of the values; the smaller it is, the lower the spread of values around the median.

Calculating the index of variability

The formula for the index of variability is:

$$\frac{\frac{1}{2}\text{ inter-quartile range}}{\text{median}} \times 100$$

The index of variability for area A is therefore:

$$\frac{7025}{77\,500} \times 100 = 9.1\%$$

The index of variability for area B is therefore:

$$\frac{1225}{53\,350} \times 100 = 2.3\%$$

The higher the index, the wider is the spread of values around the median. Conversely the lower the index, the closer the spread of values around the median. What the statistics in this example tell us is that prices for terraced houses are more varied in area A than in area B. Your next task would be to investigate possible geographical explanations for this.

Mean and standard deviation

The most mathematically correct way is to use the mean for the average and to calculate standard deviation to show the spread of values.

Calculating the mean

The **mean** is obtained by adding up all the values and dividing by the total number.

Therefore the mean for area A is £78 810 and for area B is £53 270.

Calculating the standard deviation

The calculations for working out the **standard deviation** in area A are laid out in Figure 4.23.

For the spread of the values around the mean, the standard deviation must be related to the mean value. In area A, the mean house price is £78 810 and the standard deviation is £8360. The meaning of this is that there is a 68% probability that the price of a three-bedroom semi-detached houses in area A will be ± £8360 from the mean value of £78 810, i.e. there is a 68% probability that this type of house will be priced between £70 450 and £87 170 in area A.

In area B, the mean house price is £53 270 and the standard deviation is £2360. What this means is that there

STANDARD DEVIATION

The stages in the calculation of the standard deviation are:

1 Calculate the difference (d) in price between each house and the mean. For example, in area A the house price is £72 500; the mean is £78 810 and the difference is £6310.

2 Each of the differences are squared (d^2). For example, $6310 \times 6310 = 39\,816\,100$.

3 Having done these for all 15 house prices in area A, the 15 d^2 values are added together to give the total (Σd^2).

4 This total is then divided by 15, the number in the study ($\Sigma d^2 n$).

5 The square root of this value is taken to give the standard deviation.

Using this formula, the standard deviation for area A works out to be 8360, and for area B works out to be 2360.

Figure 4.23 Calculations for the standard deviation of house prices in area A

is a 68% probability that the price of a three-bedroom semi-detached houses in area B will be ± £2360 from the mean value of £53 270, i.e. there is a 68% probability that this type of house will be priced between £50 910 and £55 630 in area B.

TASKS

1 On graph paper, draw a dispersion diagram to show the house prices in Figure 4.20.

2 Describe what Figure 4.22 shows about the differences in house prices between areas A and B.

3 Think of two other types of data, other than house prices, that a dispersion diagram could be used to show.

Showing and testing relationships

If you have collected two sets of data, you may expect that a relationship exists between the two sets. Figure 4.24 gives examples of data sets between which a relationship can be expected.

In Figure 4.24, all the items in data set 1 are examples of what are called dependent variables; they depend upon the value of the independent items in data set 2.

Data set 1	Data set 2	Relationship expected
Temperature	Height of the land	Negative – a fall in temperature with greater height
Depth of soil	Angle of the slope	Negative – less deep soil with increasing slope steepness
Size of pebbles	Distance from river source	Negative – smaller pebble with source greater distance from river source
Time taken to work	Distance between the home and work	Positive – greater time taken with increasing distance travelled
Size of pebbles along beach	Distance from the headland	May be positive or a negative depending on the direction of drift

Figure 4.24 Relationships between data sets

Scatter graphs

Scatter graphs are used to show the relationship between two sets of data.

How to draw a scatter graph

- Draw two straight line axes.
- Label your axes so that
  - The dependent variable is on the horizontal (*y*) axis
  - The independent variable is on the vertical (*x*) axis (the only exception is height which goes on the vertical axis).
- Choose suitable scales to cover the range of values.
- Plot by dots or crosses the points at which there is a relationship.
- Do not join up the dots.

What do scatter graphs show?

Figure 4.26 shows the three main types of relationships you can expect.

- Figure 4.26(a) shows positive correlation. The relationship is positive because, as one increases (distance from place of work) so too does the other (travelling time).
- Figure 4.26(b) shows negative correlation. The relationship is negative because one decreases (pebble size) while the other increases (distance from the river's source).
- Figure 4.26(c) shows no correlation. It is not possible to detect any consistent relationship between wealth of people and frequency of visit to the local sports centre. When a strong relationship exists, as in 4.26(a) and 4.26(b), a best-fit line can be drawn in, which summarises most of the relationships shown.

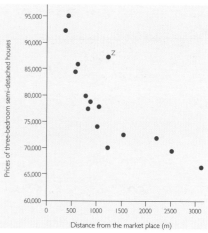

Figure 4.25 Scatter graph to show the relationship between house price and distance from the market place in a British town

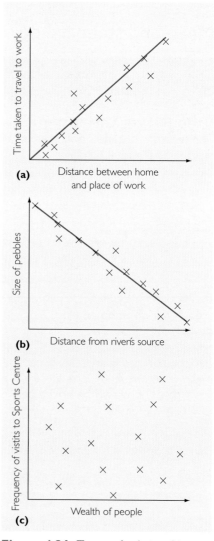

Figure 4.26 Types of relationships shown on scatter graphs

Figure 4.25 shows a negative correlation between house price and distance from the market place in a British market town. This is the opposite of what might be expected; in most British towns house prices increase with distance from the centre. There are always exceptions to any general rule. This example illustrates how useful the scatter graph can be for showing trends in the data. In addition, exceptions to the general trend stand out, such as the house marked Z on Figure 4.25. Such exceptions may be worth special study and explanation.

The Spearman's Rank Correlation Coefficient

Always draw a scatter graph first because it will give you an idea of the *nature* of the relationship. Calculating the correlation coefficient gives you the *strength* of the relationship. Figure 4.27 shows you the calculations for working out the value for the **Spearman's Rank Correlation Coefficient** for the same data used in drawing the scatter graph in Figure 4.25.

How to calculate it

- List your two sets of data in two columns side by side.
- Rank both data sets (in Figure 4.27, house prices are ranked from highest to lowest and distances are ranked from furthest to nearest).
- Work out the difference (d) in rank between columns one and two.
- When two or more values share the same rank (two houses from Figure 4.27 are 1200 metres away), the middle of the tied values are used for each one. In this example, ranks 5 and 6 are shared between the two to give each one a rank of 5.5.)
- Square each of the differences d^2.
- Total up the differences Σd^2.
- Put them into the formula for Spearman's Rank in which n is the number of data sets that were used.

$$Rs = 1 - \frac{6\,\Sigma d^2}{n^3 - n}$$

House prices (£)	Rank from highest to lowest price	Distance from marketplace (M)	Rank from furthest away to nearest	Difference between ranks (d)	Difference in rank squared (d^2)
95 000	1	400	14	13	169
92 250	2	300	15	13	169
87 500	3	1200	5.5	2.5	6.25
85 950	4	600	12	8	64
84 500	5	550	13	8	64
79 950	6	750	11	5	25
78 950	7	850	9	2	4
77 500	8	800	10	1	1
75 900	9	1000	7	2	4
73 950	10	950	8	2	4
72 500	11	1500	4	7	49
71 900	12	2200	3	9	81
70 000	13	1200	5.5	7.5	56.25
69 500	14	2500	2	12	144
66 500	15	3100	1	14	196
					$\Sigma d^2 = 1036.5$

$Rs = 1 - \dfrac{6\,\Sigma d^2}{n^3 - n}$ $= 1 - \dfrac{6 \times 1036.5}{15^3 - 15} = \dfrac{6219.0}{3360}$ $= 1 - 1.85$

$= \mathbf{-0.85}$

Figure 4.27 Calculations for the Spearman's Rank Correlation Coefficient

Advice on presenting data

Writing about what your maps, tables and graphs show is the first stage in analysing the data, which will be dealt with in the next chapter.

Variety and suitability

Variety is the key word. Set out with the intention of displaying different types of data in as many ways as possible. This is where good data collection helps. Choose those methods best suited both to the data and to the comments you wish to make towards the aims of your study.

Often there is more than one way to present the same data. In a survey of two housing areas in a city, a student discovered the following data.

Look at Figure 4.28. This data is shown using three different methods – pie graph, divided bar graph and pictograph. What is the fourth method that has been mentioned in this chapter that can also be used?

Type of housing	Area A	Area B
Terraced	40%	15%
Flats	35%	5%
Semi-detached	23%	70%
Detached	2%	10%

Which one is best to use? It depends what you want to emphasise. If showing the relative size of the component parts is most important, the pie graph is likely to be the best. If being able to see actual percentages is more important, the divided bar may be better. However, if visual representation (what it looks like) is considered to be more important than accuracy, the pictograph could be the best choice.

Do not use a method that is unsuitable for showing the data just because you feel that you must include one example of that technique to demonstrate that you know how to use it. You cannot be expected to include an example of every method in your coursework. You are only expected to use a variety of appropriate methods.

TOP TIPS

- Seek to include at least six different methods.
- Eight or more is better.
- Keep a checklist.

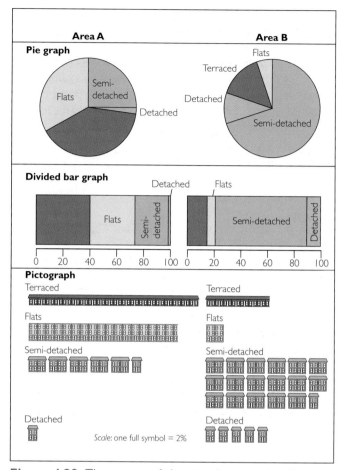

Figure 4.28 Three ways of showing the same data for housing types

Accuracy

The need for accuracy should speak for itself. A map on which towns are plotted in the wrong places is no good to anyone. The best motto is – whatever you do, do it well! It is better to draw a small number of accurate graphs and maps than many that are incomplete, inaccurate and rushed.

For neatness alone, there are usually relatively few marks, but neatness is a great help in increasing the effectiveness of your maps and graphs, for which there *are* marks. Coursework moderators are no different from other people and like to see neatly presented graphs, maps and diagrams. Therefore, neatness always helps because it has a subconscious effect on the person marking your work.

Check that everything is complete before you hand your work in. To be considered complete, all maps, tables, graphs, etc., should have a title and be numbered. Most will need a key and labels as well.

Originality and complexity

In some studies these may not be needed. Keeping everything simple is often good policy. However, if you are hoping to achieve a high grade for your coursework, there is likely to be a requirement that you demonstrate that you can use more complex techniques (like the ones used on pages 40–43, for example).

Remember that you are not going to be given extra credit just for including a more complex type of graph, such as a triangular graph, if its use is inappropriate. Unless you have data for three components adding up to 100%, there is no way a triangular graph can be used in your work. The data from which the triangular graph in Figure 4.29 was drawn

was collected by a student from observations in three quite small areas of the city. If a fourth type of housing had been observed, such as flats (as in Figure 4.28), the student would not have been able to use the triangular graph.

Integration

When the main reason for presenting data is to help understanding of what the data shows, it is therefore vital that you refer to your maps, tables, photographs and graphs. Never be afraid to add labels to any of them to highlight their main features. Labels do not ruin the appearance of photographs and graphs. Instead they tell the moderator that you recognise what is important. All Figures should be placed as close as possible to the written work that they refer to. Resist the temptation to have a part that is all graphs, or to include a block of ten pages of photographs. You may think that this makes your work look more attractive, but there is a great danger that it will look more like a photographic album than a piece of geographical coursework.

Survey of 100 houses in each of three areas

Area X		Area Y		Area Z	
Terraced	15	Terraced	80	Terraced	8
Semi-detached	75	Semi-detached	18	Semi-detached	61
Detached	10	Detached	2	Detached	31

Figure 4.29 Triangular graph of housing types in three areas of a British city

Look ahead to pages 72–73 for an example of a top grade data presentation from a student.

Chapter 5
Writing it up

What needs to be done?

You are advised to write up and present your work in four main sections.

1 **Introduction** (see pages 54–57). What were my aims and where did I do the work?
2 **Data collection** (see page 58). What did I do?
3 **Analysis of the results** (see pages 59–61). What did I find out?
4 **Conclusion** (see pages 62–63). How well did my conclusions fit the aims of the work?

You must be organised. The eight pages and section headings, which are essential for making sure that your coursework is complete, are given in Figure 5.01.

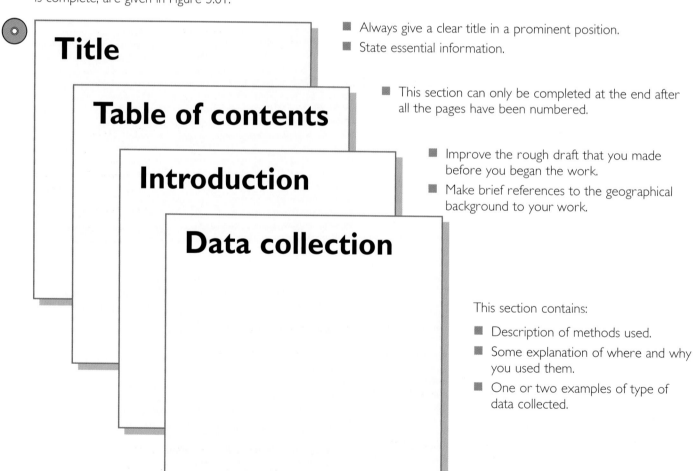

Title

- Always give a clear title in a prominent position.
- State essential information.

Table of contents

- This section can only be completed at the end after all the pages have been numbered.

Introduction

- Improve the rough draft that you made before you began the work.
- Make brief references to the geographical background to your work.

Data collection

This section contains:

- Description of methods used.
- Some explanation of where and why you used them.
- One or two examples of type of data collected.

Analysis of the results

- Present the data.
- State what it shows.
- Look for similarities, differences, patterns, links and relationships for the data.

Conclusion

- Take an overall view of the work.
- Summarise the work done and what it shows in relation to the original aims.
- Relate your findings to their geographical background – mention strengths and limitations.

References

This section lists:

- References used, for example, books, newspapers.
- People consulted.
- Any other data sources such as CD-Rom.
- Acknowledgement of software/programs used.

Appendices

This final section contains:

- Data collection sheets.
- Completed questionnaires.
- Tables of data.

Figure 5.01 Page and section headings for writing up coursework

Some of these sections are likely to be just one page long, such as *Title, Table of Contents* and *References*, but they are still important in finishing off the task properly. The longest section should be *Analysis of the results*. Putting all your work together and writing it up is your most challenging task. Try to make it as easy as possible – set your stall out before you begin.

Before you start

Assemble:

- Your data collection sheets
- The rough draft you did for the introduction
- All your maps, tables, graphs and photographs
- Any jottings or notes you made about them

Prepare your section headings as suggested in Figure 5.01.

Writing the Introduction

The *Introduction* is really important.

There are two reasons for its great importance. Firstly, it is the first piece of written work that the person marking your coursework is going to read. You need to do all you can to make a favourable impression. Secondly, a good introduction sets you up for the rest of the study. It is the foundation upon which the rest of your written work is based.

What should be included?

1 The title and aims (purpose) of the study.
2 Some information about the study area and its suitability for your study.
3 Brief information (under one side of paper) about the geographical background to your topic.

Title and aims of the study

Have another look at what you suggested originally for the title and aims of the coursework. Do they accurately reflect the work you have now done? This is your chance to change them if they are not a good fit for the data collected. A good title is vital, because it is so much easier to write about what you have found out and to conclude well.

Information about the study area

Maps are essential for showing the general location of the study area and the main sites where you did the fieldwork. If you look ahead to pages 70–71 you will see a good example of a student using appropriate maps in a river study. One map shows the whole of the river's drainage basin; the second map is on a larger scale and shows sites where channel measurements were made.

Whatever your topic, include at least two maps at different scales. For example, let us imagine that you are examining visitor pressure at a honeypot site in a National Park. Begin with a map of the National Park on which the location of the honeypot site is named and highlighted. Use a second map to give more detail of your study area around the honeypot.

Many students automatically include a copy of part or all of a printed map in the *Introduction*. Printed maps contain so much information that it is often difficult to highlight study sites so that they stand out clearly. By tracing an outline, and drawing one of your maps as a sketch map, you can concentrate on showing locations that are relevant to your study.

Give some background information about the study area. The type of information will depend upon the theme and aims of your study. Information may be:

■ Physical – the geology, relief, climate or soils of the area
■ Economic – employment and economic activities such as farming, forestry, mining, industry, tourism and other service industries
■ Social – population details
■ Historical – growth and change in the settlement.

Whilst background information is important, the rule is – **keep it brief and to the point**. Page after page of background information is not needed. You will be wasting your time. In particular, students undertaking a town or village study often give far too much historical information, having forgotten that they are supposed to be doing geographical (and not historical) coursework – see Figure 5.02. One way to avoid writing down too much background information is to summarise it in a table. This will help you to select the key points and limit the amount. Historical information can be put down in a time column or time line as in Figure 5.03a and b.

Figure 5.02 Be geographical

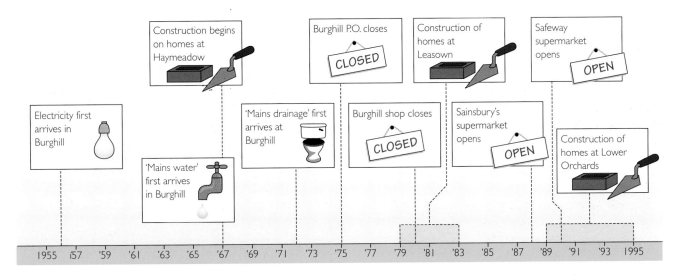

Figure 5.03a Example of a time line

Date	Information/Change
1750	Farming village; houses grouped around church
1901	Coal mine opened; village grew; terraced houses built around the mine
1905	Railway station opened
1952	Council housing estate built
1963	Railway station closed
1967	Mine closed
1982	Private housing estate built

Figure 5.03b Example of a time column

Maps need not be the only type of illustration used in your *Introduction*. In some studies, one or two (but not too many) photographs may be useful for showing features of the study area, provided that you label them. If you are dealing with a local issue, the headlines from local newspapers, cut out and stuck in, can have a big impact.

Geographical background

Briefly give information about the geographical background for your study. You can find the information in a textbook, although often you will have covered what you need in geography lessons. Below are some examples of information for the geographical background.

- For a river study, you could refer to the expected changes in a river and its valley between source and mouth. You may make brief mention of the processes responsible.

- For a study of the CBD, you could mention the typical features of a city centre that make it different from the other urban zones.

- For the study of the honeypot site in a National Park, you could either give some background information about National Parks or you could mention the general problems caused by visitor pressure. Which one of these is more appropriate for your study will depend upon your aims.

TOP TIPS

- Keep the geographical background brief – one side of A4 paper is usually sufficient.
- Choose the geography most relevant to your aims.
- Don't copy out large amounts of information from books.

DOs AND DON'Ts

DO

✔ Start to number the maps and other illustrations
e.g. Figure 1 and Figure 2

✔ Try to include a hand-drawn sketch map

DON'T

✗ Begin with a map of the world, followed by a map of the UK

✗ Use a photocopied map without anything added by you

Introduction – examples of students' work

Extracts from two *Introductions* are given below. The purpose is to guide you about what you should, and should not do, when writing up your own coursework. Student A's title was 'Does Bicester have enough recreational facilities for the whole community?' Student B's title was 'Does the Arnison Centre conform to the typical characteristics of an out-of-town shopping centre?'

Work of Student A

Aim
My aim is to target every age group to ascertain if they are satisfied with the amount of recreational facilities provided for them. If they are not, I want to establish how the facilities could be improved and if this would encourage more uses of them in Bicester.

Planning
I intend to give out questionnaires in Bicester Town Centre and ask ten people from the following age groups: 10–16: 17–25: 26–47: 48–60: and 60+.

Predictions
I believe that older people will be satisfied with the recreational facilities in Bicester because there are a number of facilities, for example, public houses, bowling clubs and community centres. Whereas most people between 14–25 would like more facilities. I feel that Bicester needs a bowling alley, cinemas and a nightclub or disco. Although this would be very expensive, I maintain this would be very beneficial to the community.

Problems
My main problem is the quality of the answers to the questionnaires, and the level of response to my letters.

Figure 5.04 Student A's *Introduction*

Introduction Checklist

Title indicated	✗
Aims of study	✓
Study area	✗
Geographical background	✗

CHIEF MODERATOR COMMENTS

○ This student failed to plan what to include in the *Introduction*.

○ As a result, the student neglected three out of the four items on the checklist.

○ Anyone reading it is not going to be impressed – a real problem for the student.

○ Problems are better mentioned later in the *Data collection* or *Conclusion* sections rather than in the *Introduction*.

TASKS

1 Make and complete a checklist as (below) for Student B's *Introduction*.

2 Look at Figure 5.06. Do all of these questions help Student B to answer the principal aim stated in line one? Do you think a smaller number would have been better? Explain your answers.

3 Student B word-processed the work – but can you find the two mistakes that defeated the spell check?

Work of Student B

Figure 5.05 Arnison Shopping Centre

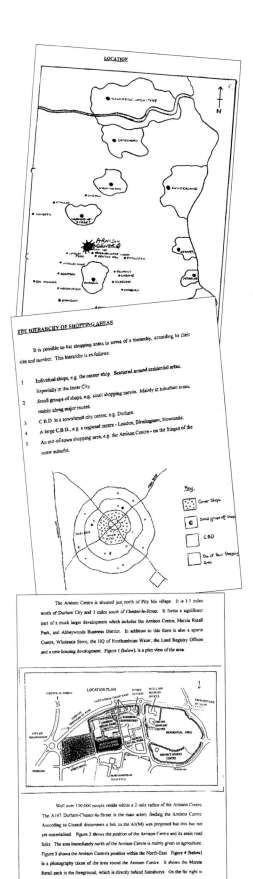

Does the Arnison Centre conform to the typical characteristics of an out-of-town shopping centre?

Part A. Page 1.

The principle aim of this fieldwork project is to establish whether the Arnison Shopping Centre conforms to the typical characteristics of an out-of-town shopping area. In addition to this, the data collected during the course of the fieldwork will help to answer the following questions:

1 What is the sphere of influence of the Arnison Centre?
This is to discover how far people are prepared to travel to shop at the Arnison Centre.

2 What shops and services are available at the Arnison Centre?
This is to discover how the Arnison centre caters for the public, and what types of goods it sells.

3 Which shop is the most popular?
The aim is to find the most popular shop and find out why it is so popular.

4 How do most people travel to the Arnison Centre?
This is to discover the most popular form of transport.

5 Is the Arnison Centre easily accessible?
This is to find out if there are enough roads to cater for the people wanting to visit the Centre.

6 Are there adequate parking facilities for able and disabled people?
Are there enough disabled parking spaces near the shops and is there enough space for able bodied people?

7 What are people's views on out-of-town shopping areas?
This is to discover whether people agree with out-of-town development, taking away business from the C.B.D. (Central Business District) or not.

8 Why was the sight of the Arnison Centre chosen?
Why did the developers choose the Arnison Centre's site for development.

Figure 5.06 First page of the *Introduction* from Student B to show how it was organised

Figure 5.07a–c Extracts from the rest of Student B's coursework

Writing the Data collection and Analysis

Data Collection

The main purpose of this section is to show **how**, **where**, **when** and **why** you collected your data.

1 **How and why?** Describe your methods of data collection. Explain why you used them and why they were suitable for your study.

2 **Where and why?** State the places where you collected data. Explain why these places were suitable collection points for your study.

3 **When and why?** State the times and dates when you collected data. Explain their importance to your study.

How

Mention all the methods of primary data collection, as well as any secondary sources you may have used. It is best to start by drawing up a summary list of the methods used, similar to the one shown in Figure 5.08, so that none are missed out.

Describe each method in turn, preferably dealing with them in order of importance.

■ If you took measurements, describe the equipment used. Include one copy of a blank record sheet and one copy with some results on it.

■ If you relied upon questionnaires, enclose one blank copy of your questionnaire and comment upon the questions used; also include just one example of a completed questionnaire.

■ If you undertook an environmental quality survey, name and explain the factors that you used as the basis for doing it. It may be a good idea to use labelled photographs or field sketches to illustrate high and low environmental values.

Irrespective of the methods of data collection used, what you should be doing is giving evidence that you have collected the data; but there is no need to include all your fieldwork records in the work, even though you must not destroy them in case they are needed to prove later that your results were genuine.

Where

It is important to show that some thought went into choosing sites that were suitable for collecting your data. When doing a survey for a physical geography topic, such as along a beach or up a valley side, or for a human topic such as a land use transect through a town, the measuring and recording points are usually spaced at equal intervals. However, other factors, such as safety or access problems, might have stopped you from doing this and these should be explained here. You have much more control over choice of sites when taking pedestrian or traffic counts or handing out questionnaires. Explain why you considered the sites chosen to be the best ones for your data collection.

When

The times at which the fieldwork was undertaken, or the length of time spent counting and observing, can be very significant for the collection of accurate and reliable results.

PRIMARY SOURCES

Questionnaire:
■ Questions to shoppers

Observation:
■ Land use survey
■ Environmental quality survey

SECONDARY SOURCES

■ Letter to Council
■ Newspaper cuttings from Reference Library
■ Shopping centre's website

Figure 5.08 An example summary list of data collection methods

Ready to start writing the Analysis?

The Analysis a mixture of written text and illustrations.

- Have a sheet of paper with the title and your aims on it in large type.
- Keep it next to you so that you don't forget what you are supposed to be writing about.
- Remember that you should be looking to place maps, graphs, photographs, etc., as close as possible to the text to which they refer.

TOP TIPS

To write an *Analysis* of Grade A standard:

- Demonstrate your skills of selection. Don't try to write about every feature on every Figure. Concentrate upon those that are significant for the aims of your study.
- Demonstrate higher level skills of analysis. Spend more time commenting on differences, patterns and relationships.
- Be willing to do a little extra/new data collection should you find yourself short of information in one part of the study when writing up the work.

Figure 5.09 Ready to start

Starting to write

Work from the easy towards the more difficult.

Stage 1

- Begin by making simple and straightforward statements about what each Figure shows.
- Quote numbers, values or other evidence to support your statements.
- Try to identify the most significant information, such as the largest, the smallest, or the one showing greatest change.

Stage 2

Then add more detail and explain.

- Give (or try to give) reasons for what each Figure shows.

Stage 3

Extend the analysis (particularly if you are hoping for a high grade).

- Try to identify similarities, differences, patterns, links and relationships.
- Look at X, Y and Z below for ideas, because what you will be able to do depends a lot upon the nature of data collected.

X For data from different places or data collected at different times:
  - How are they different?
  - In what ways are they the same or similar?
  - Why do these differences/similarities exist?

Y For mapped data:
  - What patterns or variations are shown?
  - Why have they developed?

Z For an analysis of two sets of data:
  - In what ways are the two sets of data related?
  - How strong is the relationship?
  - Why do data links occur?

Analysis – examples of student work

Student A

Pages 7–9 Tally charts of replies from the questionnaires.

Page 16 Pictogram showing the numbers using each leisure facility.

Page 18 Desire lines with distances to show how far people travelled.

Pages 10–15 Bar graphs for showing results from the questionnaires.

Page 17 Pie graph for reasons why people use facilities outside Bicester

Pages 19–21 Written part of the Analysis.

Figure 5.10 Summary of Student A's Analysis

Student A's written analysis began on Page 19 with:

Graph 1 (a) What facilities do you use in Bicester?

This graph shows that Bicester Sports Centre and the Public Houses in Bicester are the most popular facilities. Most people aged between ten and sixteen use the Sports Centre. Most people between 17 and 25 use public houses in Bicester. The most popular facility for people aged between 26 and 47 are public houses as well. The most popular facility for people aged 60 are clubs. I think that the public houses are so popular because there are so many of them and the Sports Centre is so popular because it is the main attraction at Bicester and it also holds many different sports so that it has something that everyone likes.

Graph 2(a) How often do you use these facilities?

CHIEF MODERATOR COMMENTS
- One or two values have been used to support the simple statements made.
- These follow the word 'because', but the reasons have been stated in only a very limited way.
- Each Figure has been looked at separately by the student with no attempt to relate one Figure to another.
- The Figures have been included in a block separate from writing. Graph 1(a) was drawn on page 10 and the student wrote about it on page 19.

CHIEF MODERATOR COMMENTS
- Quite a variety of presentation techniques were used, but it is a pity that the student didn't make best use of them.

MODERATOR'S OVERALL COMMENTS

For data presentation – Level 2 standard

For analysis – Level 1 standard (for levels marking see pages 104–5).

Checklist

Has the student described what each Figure shows? (See Stage 1 page 59)	✔
Has the student begun to give reasons for what is shown? (See Stage 2 page 59)	✔
Has the Analysis been extended in any way? (See Stage 3 page 59)	✘
Has the Analysis been organised correctly?	✘

Tackling Geography Coursework

RESULTS – QUESTIONNAIRE

The questionnaire was taken on two different days – Wednesday and Saturday. In some of the questions, the results are presented together and merged. Sometimes they are presented separately when it is relevant, because they will point out a pattern.

I Where have you travelled from to shop at the Arnison Centre?
The results of this question are presented separately in Figures 13 and 14.

Figure 13
The results of the questionnaire, taken on the Saturday morning, show that the Arnison Centre has a large sphere of influence, around 17km. The majority of shoppers, however, travelled only up to 8km to shop at the Arnison Centre. The largest number of people who were questioned came from within Durham City, which was a journey of less than 3km. Although the city centre of Durham has a much larger choice of shops than the Arnison Centre, because it is a higher order shopping centre, parking is free at the Arnison Centre.

Figure 14
The results of the questionnaire taken on the Wednesday evening show that during the week the sphere of influence of the Arnison Centre decreases by about 7km. Like the Saturday questionnaire, most people visited from Durham City. But this may be attributed to people returning home from work via the A167 stopping off to buy food or to get petrol. Unlike Figure 13 there is not such a distinctive pattern in the way in which the number of shoppers decreases as the distance increases.

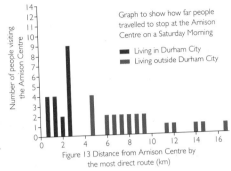

Graph to show how far people travelled to stop at the Arnison Centre on a Saturday Morning

■ Living in Durham City
■ Living outside Durham City

Figure 13 Distance from Arnison Centre by the most direct route (km)

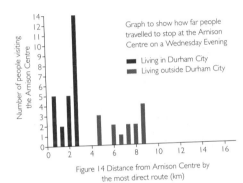

Graph to show how far people travelled to stop at the Arnison Centre on a Wednesday Evening

■ Living in Durham City
■ Living outside Durham City

Figure 14 Distance from Arnison Centre by the most direct route (km)

Figure 5.11 First part of Student B's *Analysis*

CHIEF MODERATOR COMMENTS

○ Several descriptions from the Figures have been made.

○ Reasons have also been given for these.

○ The student is looking for comparisons between the data collected on the two different days. Some of the similarities and differences between Figures 13 and 14 have been identified. Also the student has started to suggest reasons for them.

○ Figures 13 and 14 have been placed on top of each other so that it is easy to compare them. These Figures were placed next to the written part to which they referred.

MODERATOR'S OVERALL COMMENTS

○ High Level 3 performance throughout

Checklist

Has the student described what each Figure shows?	✔
Has the student begun to give reasons for what is shown?	✔
Has the analysis been extended in any way?	✔
Has the analysis been organised correctly?	✔

TASKS

I From Student B's written analysis identify examples of **(a)** descriptions of what the Figures show **(b)** reasons and **(c)** comparisons.

2 Make a list of all the things that make the analysis of Student B worth more marks than that of Student A.

Conclusion

It is essential that you have a section in your coursework headed *Conclusion*.

The main purposes of this section are:

- To look at all the work done.
- To link the results to the original aims of the work.
- To draw overall conclusions.
- To place your work in its broader geographical setting.
- To evaluate the strengths and weaknesses of the work as a whole.

Figure 5.12 below is a flow diagram to take you through the stages for writing your conclusion.

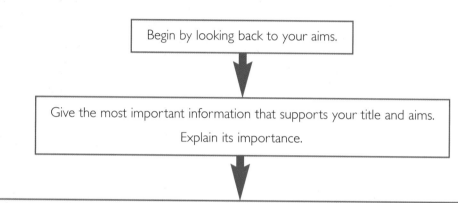

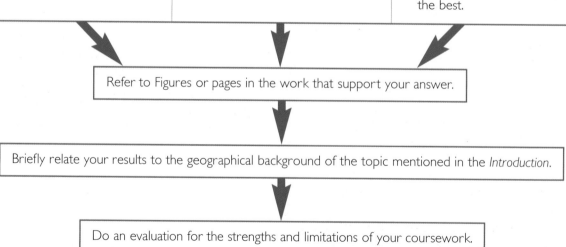

Figure 5.12 How to write up the Conclusion

Tackling Geography Coursework

Conclusion – examples of a student's work

In Figures 5.13–5.14 below are four extracts from Student B's *Conclusion*.

A. Opening statements for the Conclusion
From the data collected during the course of this study it is apparent that in almost every respect the Arnison Centre fulfils the essential characteristic of a 'typical' out-of-town shopping centre. It is a well-planned, pleasant environment, selling goods to cater for many needs. It has links to major roads and is built upon what was farmland. It has large parking facilities. Some of these 'typical' features are shown on the Photograph.

B. An extract from the next part of the Conclusion
The study also sought to consider a number of supplementary questions. The answers to these have been summarised below.

1 The sphere of influence of the Arnison Centre differs depending on the day visited. During the week it is around 10 km but at the weekend it increases to around 17 km, due to the fact that Saturday is peak shopping time

2 The Arnison Centre provides shops which…

C. An extract from a later part in the Conclusion
The Arnison Centre development is only a few years old. Currently this type of centre is the source of some fairly heated debate, in that only now are local planners beginning to consider some of the negative aspects of such centres. Perhaps a logical follow up to this study would be to consider some of the negative effects its presence has had on trade and commerce in Durham City itself. Looking at where people travel from to shop at the Arnison Centre it is fairly clear that the centre must have had some detrimental effect on the level of business activity in Durham City. The popularity of out-of-town shopping centres must give planners some food for thought when they come to consider their further development. Perhaps the next ten years will see town and city centres fighting back as they incorporate some of the characteristics of out-of-town shopping centres into their own redevelopment.

D. The conclusion was rounded off with Limitations
The number of people sampled in relation to the number of people using the Arnison Centre is small and therefore may not be representative of the majority of the people. If a questionnaire had been used on every day of the week at a certain time, then this would have provided a more representative sample. The Arnison centre is part of a much wider retail park complex. To study just one part of this complex has its limitations.

Figure 5.13–5.14 Four extracts from Student B's Conclusion

All eight of Student B's introductory questions were answered in the *Conclusion*.

Checklist

Results linked back to original aims – Conclusion for the main aim and for the eight smaller aims ✔

Overall conclusion drawn – Begins with an answer to the question in the title. Photograph adds impact. ✔

Work placed within broader geographical setting – References made to out-of-town shopping centres. ✔

MODERATOR'S OVERALL COMMENTS

Confirmation of a top Level 3 performance – of grade A* standard

Finishing off your work

Don't be too worried if your data does not lead to a clear-cut answer. For example, in a river study, while it is reasonable to expect that a river as it flows from source to mouth will become wider and deeper, and that it will have a load of smaller sized particles, one or more of the study sites may not fit the expected pattern. All you can do is to suggest reasons for the difference based upon your observations at the site. Human interference is one possible explanation.

Similarly in urban studies, a clear dividing line can exist between one urban zone and the next. The shops and offices of the CBD may be suddenly replaced by houses along one of the urban transects you are using. However, along another transect route you may find a mixed area of shops and houses, which makes it impossible to draw a precise dividing line where the CBD ends. All you can do is to attempt to explain why. Comment about the exceptions could form part of your evaluation.

The conclusion is not normally the part of the coursework in which you look to present maps, graphs, photographs or sketches. Usually it is all writing. However, if you can find some kind of illustration that is relevant in the conclusion, its use may have great impact and help to round the work off well. Some suggestions are offered in Figure 5.15, but a lot depends on the nature of the study. If you have a chance to show initiative, take it and end on a high note.

■ A flow diagram showing the sequence of steps that led to your conclusion; perhaps you can extend it further by adding future possibilities.

■ A time line, as illustrated on page 55, which summarises the main changes and when they occurred.

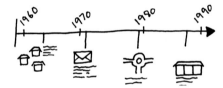

■ A map, which you might have used earlier, but on which you add new labels to highlight your main conclusions.

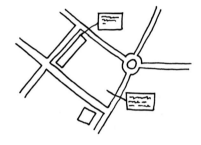

■ A summary chart giving arguments for and against a proposal; perhaps you can add to it some indication of what you consider to be the relative strengths of the arguments.

■ Pictorial drawings summarising the main differences between the places or areas you have used in your work.

■ A new photograph on which some of the summary features are labelled or one which makes a neat contrast to the one you used earlier, for instance in the *Introduction*.

Figure 5.15 Suggestions for illustrations to round off your *Conclusion*

Writing an evaluation

Including an evaluation is like putting the icing on a cake – the cake is more important than the icing, but the evaluation shows that you have been willing to take an overview of the entire coursework process. You should refer to one or more of the following as appropriate for your work.

Strengths such as

- What you found most interesting or useful or rewarding in doing the work
- What you consider to be the strengths and successes of the work.

Weaknesses and limitations such as

- Problems or difficulties encountered in collecting the data
- Whether it would have been better to have collected more data or data of a different type
- How you might be able to reduce or overcome these weaknesses and limitations if you were to do another similar study.

Wider applications such as

- You have done a 'snap-shot' study of one example; is your example typical? How likely is it that your conclusions could be applied to other examples?
- Possibilities for doing further work or for extending the study.

In the majority of cases, one side of A4 will be more than sufficient for the evaluation. That gives you enough space to demonstrate that you have thought about what you were doing and that you have benefited from the experience of doing the coursework. Problems for data collection always exist where there is water (Figure 5.16). Questionnaires can provide a lot of valuable information but can lead to a host of problems (Figure 5.17). Organisations and companies do not always reply to letters. If they do, they often won't answer the questions you asked them; instead many just include their printed publicity materials. If you made use of environmental quality surveys you could explain their subjectivity, i.e. another person's assessment could be very different to that of your own.

- People refusing to answer questions (too busy!)
- People not filling in questionnaires delivered to the house (the dog ate it!)
- Non-performing questions (questions not giving you the information you really wanted)
- Too small a sample (more questionnaires needed)

Figure 5.17 Questionnaire problems

DOs AND DON'Ts

DO

✔ Be as positive as possible

✔ Sell your work to the moderator

DON'T

✘ Be too self-critical

✘ Make long lists of problems and inadequacies

And finally ...

There are the routine tasks and checks to make sure that all is finished.

- Fill in the page with the title 'References'. Only a list is needed – for websites, books, newspapers, maps, videos, etc., used by you. Name any people or organisations who were contacted by you and from whom you received information useful to your coursework.

- Add the page numbers and check that you have changed the numbers on the Figures from pencil to ink.

- Finish off the Table of Contents at the front by adding in the page numbers.

- Check that page and Figure references match up.

- Proof read for spelling, punctuation and grammar.

- Check everything, so that the person reading your work is going to gain the most favourable impression possible.

Figure 5.16 What can go wrong during a river study

Chapter 6
Topic examples

Tackling rivers

What can you do?

1 Measure **channel shape**
- How and why does the shape of the river channel vary from place to place?
- What are the differences in channel cross-sections between meanders and straight sections of the river?

2 Take readings for **speed of flow**
- Hypothesis – that the speed of flow of the river increases downstream.
- How and why does the river's speed of flow vary along its course?

3 Concentrate on **discharge** (volume of water passing through the channel at a particular time)
- Hypothesis – that amount of river discharge increases downstream.
- What are the differences in the river's discharge between summer and winter?

4 Measure **gradients**
- Hypothesis – that channel gradient decreases downstream.
- Is there a relationship between channel gradient and speed of flow?

5 Examine **load size** and **shape**
- Hypothesis – that load size decreases downstream.
- In what ways and why do shape and size of the load vary along the river's course?

6 Look at the impact of **people**
- How and where have people altered the course/channel of the river?
- What shows that some stretches of the river have been managed by people?
- Is there any evidence of water pollution?

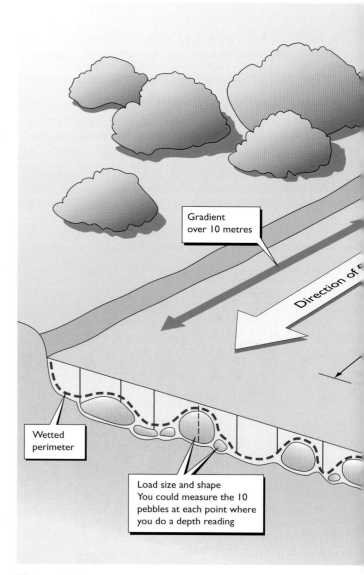

Gradient over 10 metres

Direction of

Wetted perimeter

Load size and shape
You could measure the 10 pebbles at each point where you do a depth reading

Figure 6.01 Measurements taken in a river channel

How do you do it?

Figure 6.01 shows some of the measurements that can be made at each site along a stream.

Figure 6.02 shows students engaged in measuring the **channel shape**. Measure the width at the water surface using a tape or rope. At the same time measure the depth at various points across the river. This may be at 25 cm intervals or you could take ten evenly spaced readings across the channel.

Extra information on channel shape is given by measuring the **wetted perimeter**. This is the distance with which the water is in contact. Use a rope or long piece of chain and measure the stream bed. You need to go up and down around rocks and stones. Measure it at the same place as you measure the width. Hope for a warm day because British river water is always cold. Have a small towel and be prepared to wear gloves, even in mid-summer!

Figure 6.02 Measuring water depth and channel width

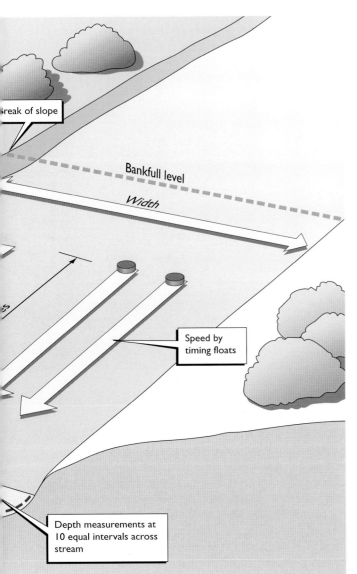

reak of slope

Bankfull level

Width

Speed by timing floats

Depth measurements at 10 equal intervals across stream

Figure 6.03 Measuring speed with a flow meter

To measure surface **speed of flow**, you have a choice of two methods. One is with a flow meter (Figure 6.03). This records the true velocity below the surface. You should take several readings across the stream and work out the average.

The second method is more manual, but in some ways more enjoyable (Figure 6.04). Floats, such as small pieces of wood, oranges or even dog biscuits, are timed over a 10-metre stretch of river. Since floats will get caught behind stones and held up in turbulence, it is a good idea to do the measurements at a minimum of three places across the stream and to carry them out several times. Surface speed is worked out by

$$\frac{\text{distance}}{\text{time}}$$

Take the average for all of the speeds in that section of the river. Surface water flow is reduced slightly by friction with the air above. You should multiply your results by 0.85 to gain a more accurate idea of the overall speed for a small stream.

Figure 6.04 Measuring speed of flow using a float

Figure 6.05 How to measure the gradient of a stream

Once you have measured width, depth and speed of flow, you have all the data you need for calculating river **discharge**. The formula for calculating discharge is

$Q = A \times V$
where Q is the discharge (m³/sec)
A is the cross-sectional area (m²)
V is the velocity (m/sec)

For an example of this, look ahead to page 70.

Gradient is the slope of the river. It is measured using two poles and a clinometer. How this is done is explained in Figure 6.05. Figure 6.06 shows students engaged in measuring a gradient.

1 Measure a 10-metre stretch of stream.
2 Carefully hold two metre rules on the water surface 10 metres apart.
3 Use the clinometer to measure the angle between the tops of the two metre rules.
4 Do it two or three times to check your answers.
5 Record your result.

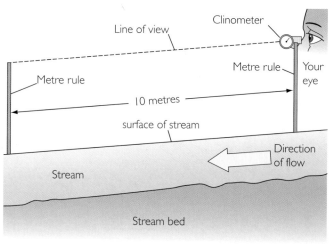

Figure 6.06 Using a clinometer to measure the gradient

Load size and shape – You would expect the rocks and pebbles on the stream bed to get smaller and more rounded as you move away from the river source, although this may not always be true if the stream passes over different rock types. To collect your pebbles for measurement you need to take a random sample of say ten pebbles across the width of the stream at each location. This could be done at the same locations that you measure stream depth. For each pebble measure the 'b' axis (width); this is the axis the pebbles will roll along in the river bed (see Figure 6.07).

Getting started – useful pieces of advice

Choosing the stream and sites to study

Most rivers are too wide and too deep for study without needing wetsuit, flippers and snorkel! If you do not know a suitable local stream, study a map of the local area, preferably a 1:25,000 map. Select several small streams before making your final choice. Remember to check that you can gain access to the stream at a sufficient number of points.

If you are studying changes downstream, you need to choose a number of different sites – six would be ideal. Try to use a recognised method of sampling (see pages 22–23) to choose your six sites, for example, you could use equal intervals along the stream. However, other factors may also determine your study sites, for example, points of access such as a bridge or footpath. You will need to explain how and why you selected your sites in your section on *Data collection*.

Figure 6.07 Measuring pebbles

Equipment checklist

- Large scale map of area of study
- Supply of recording sheets, a clipboard, notepad and heap of pencils
- Tape measure and long rope or chain
- Poles and clinometer
- Metric rule for depths and measuring pebbles
- Floats and/or flow meter
- Stop watch
- Camera and film (optional)
- Wellingtons, waterproofs and lunch.

Safety

- Wear Wellington boots.
- Do not do the fieldwork in fast-flowing areas or if the stream is in flood.
- Always have someone else with you – you will need someone to help you take the measurements.
- Do not allow yourself to get too cold.

TOP TIPS

- Choose the stream very carefully.
- Measure accurately – many river measurements will vary by only a few millimetres.
- Record the data with great care.

Figure 6.08 Be careful!

Example of a recording sheet for rivers

Site number:					Grid Reference:				
Date:	Weather:				Width:				
Slope:	Angle:				Wetted perimeter:				
Depth readings (at 10 equal intervals)	1	2	3	4	5	6	7	8	9	10
Float times (seconds) (over 10 metres)	1	2	3	4	5	6	7	8	9	10
Stone number	1	2	3	4	5	6	7	8	9	10
Length of 'b' axis										

Examples of a student's work

Short extracts from a student's coursework report are used here as a guide for writing up your own coursework.

The *Introduction*

Title: Is the River Browney a typical river?

Introduction:

	Page
Why I have decided to see if the River Browney is a typical river.	1
The geographical theory of rivers	2–4
Location of the River Browney and the Aims of the enquiry	5
Map of England showing the location of North-East England	6
Map of North-East England (showing the main towns)	7
1:25 000 Ordnance Survey maps cut up and stuck in showing	
Sites 1 – 3	8
Sites 4 – 6	9
Sites 7 – 10	10
PLANNING – the number and location of the sites	11–12

Figure 6.09 What was included in the student's *Introduction*

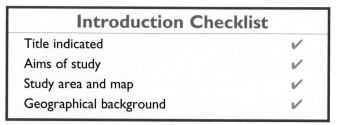

Introduction Checklist

Title indicated	✔
Aims of study	✔
Study area and map	✔
Geographical background	✔

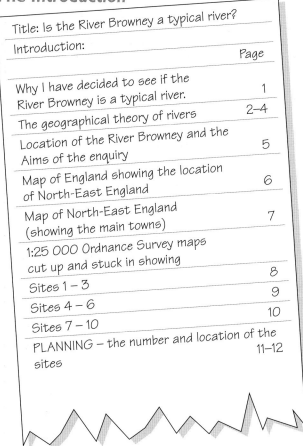

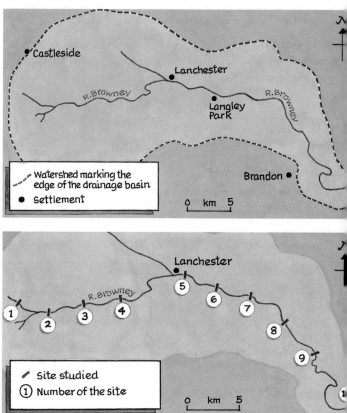

Figure 6.10 Maps showing the location of the study area

Data collection

'Number and location of sites – the river must be tested at a number of different sites along its length to decide whether the river conforms to such statements as: 'The water in the river flows more slowly as you move downstream'. The number and location of the sites that will be chosen is difficult to decide upon. The primary consideration is accessibility; it is vital that it is possible to reach the site. Another consideration is some kind of even spacing to the sites, so that each part of the river is studied equally. Finally, of course, the sites should be as much in their natural state as possible. Considering that the length of the river is 36.4 kilometres I think it would be reasonable to study it at ten sites. This means I will be able to look at the river approximately every 3.5 to 4 kilometres. My intended locations for the sites, in view of the above criteria, are the following 6 figure grid references (listed in order from nearest the source to the furthest away):

1 059447	2 096447	3 117448	4 143449
5 178464	6 228449	7 242439	8 255423
9 257409	10 267383.

Having thought about the number and location of the sites, it is now important to consider the techniques that should be used at each site to obtain the required data. Certain types of data, such as the river's width, must be collected by measurement. Other types of data, such as evidence of river deposition, must be collected by observation. Some types of data can only be collected by the use of questionnaires. People's opinions of the river countryside, for example, must be ascertained by actually asking them. There is also some data that is easier to collect from secondary sources. For example, when drawing a long profile, I will need to know the height of the river above mean sea level at a number of points along its length. It is probably more accurate to find the height by locating the point on an Ordnance Survey map. The map can also be used to draw sections for the shape of the valley.

Presenting and using the data collected

Many of the measurements taken at the study sites were placed in tables to make them easier to read (Figure 6.11).

Site	1	2	3	4	5	6	7	8
River width (m)	1.7	2.7	5.2	4.3	5.2	7.1	8.2	7.0
Depth (cm) at:								
0.5m	20	10	10	10	10	6	5	12
1m	25	15	10	10	12	7	6	10
1.5m	15	20	11	11	15	7	6	10
2m		10	12	12	17	8	5	10
2.5m		10	14	12	22	8	4	9
3m			15	10	27	8	4	10
3.5m			15	11	28	10	40	10
4m			13	10	25	15	7	11
4.5m			12		20	17	9	14
5m			10		7	20	12	16
5.5m						19	13	20
6m						19	16	23
6.5m						18	18	17
7m						14	20	0
7.5m							24	
8m							20	

Figure 6.11 River width and depth at the first eight study sites

The measurements in Figure 6.11 were used to draw channel cross-sections for each site; examples of two of these for sites 5 and 6 are shown in Figure 6.12. Notice how the student has kept the same scale to make them easy to compare.

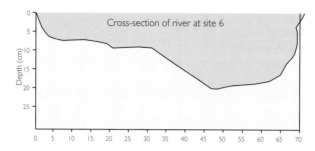

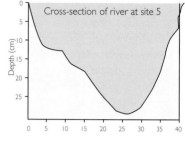

Figure 6.12 Channel cross-sections

By tracing outlines of these cross-sections on to graph paper, the student was able to work out channel area at each site. Having measured river velocity by dropping an onion into the river and measuring how long it took to travel five metres, the student was able to calculate the discharges at each site by using the formula

$$Q = A \times V$$

where Q is the discharge (m³/sec)
A is the cross-sectional area (m²)
V is the velocity (m/sec)

The student made use of secondary information to obtain height data for showing the long profile of the river by means of a line graph (Figure 6.13).

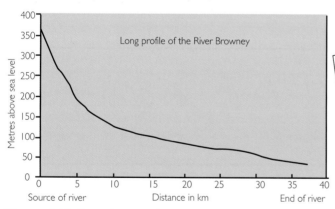

Figure 6.13 River long profile

While visiting sites to take measurements, the student observed some of the other physical features of the stream. For each site the student included a photograph (Figure 6.14). Letters and numbers placed around the edges were used to locate river and valley features identified in the student's written comments. Part of the student's text that accompanied the photograph for site 3 is given as Figure 6.15.

Figure 6.14 Photograph at site 3

'Although the river is beginning to show early signs of becoming a mature river, notice still its rocks on the bed affecting the river flow, which continues to consist of large particles (at C3), and its relatively steep banks (for example at B2), which are characteristic of a youthful river. However, there is evidence of a flood plain (at A1).'

Figure 6.15 What the student wrote about site 3

The student observed human features as well as giving out questionnaires to people using the river and river banks. The number of people recorded at each site was shown by a bar graph (Figure 6.16). A pictogram was used to show the number of items of litter found in the river and along its banks (Figure 6.17). The student used a triangular graph to show the ways in which the people interviewed for the questionnaire had reached the river side (Figure 6.18). This was an appropriate technique of presentation to use because the interviewees had reached the river by one of only three means – either by walking, or by biking or by car.

Data Presentation Checklist

Maps ✔ Photograph ✔
Tables ✔ Bar graph ✔
Cross-sections ✔ Pictogram ✔
Calculations ✔ Triangular graph ✔
Line graph ✔

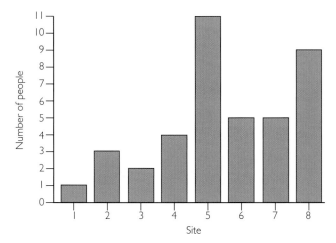

Figure 6.16 Number of people at each site

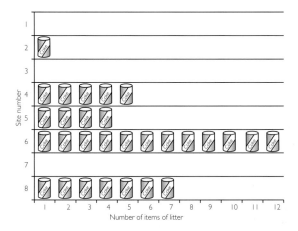

Figure 6.17 Pictogram showing number of items of litter found in the river or along its banks

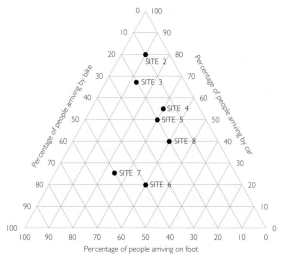

Figure 6.18 A triangular graph showing the ways in which people questioned at each site reached there

CHIEF MODERATOR COMMENTS

○ All the methods were appropriate.

○ Quite a variety of methods were used.

○ Use was made of them by the student in writing up the work.

○ Included also were some more complex techniques.

The student was able to use such a range of methods of presentation because of the amount and variety of data that had been collected. Measurement at the sites was supported by observation. Further data were gained by giving out questionnaires and by using secondary map information.

Tackling coasts

A very old song goes 'Oh I do like to be beside the seaside, I do like to be beside the sea ...' The coast is a great place for doing geography fieldwork, particularly in good weather.

What can you do?

1 Measure and observe **coastal landforms**
   - What are the main features of the beach and cliffs?
   - How and why do cliff heights and profiles change along a stretch of coast?
   - Which dominate in this section of coast – landforms of erosion or landforms of deposition?
   - Hypothesis – beach and cliff profiles become less steep towards the centre of the bay.

2 Focus on **beach studies**
   - Investigation – two or more beach transects from the low tide level to the land above high tide level to investigate similarities and differences along the beach.
   - How and why do the shape and size of pebbles change down/along the beach?
   - Hypothesis – beach profiles vary between summer and winter.

3 Examine **human influences**
   - Investigation – methods of coastal protection being used and their effectiveness.
   - In what ways and why has this coastline been managed?
   - What is the effect of the groynes upon the depth and width of beach material?
   - Why do some stretches of the coast attract more visitors than others?
   - Hypothesis – the use and economic importance of the beach to the coastal resort changes at different times of the year.

How do you do it?

Coastal areas are very varied. There are many different types of fieldwork you can do as well as many sources of secondary data you can consult. Study Figure 6.19 to find out more.

Cliff surveys

To complete a **cliff profile:**

1 Start by drawing a sketch of the cliff – you could also take a photograph which will help you draw an accurate sketch later.

2 Estimate the height of the cliff using the method shown in Figure 6.20.

PRIMARY DATA

- Measuring cliff profiles and small scale features such as caves
- Measuring the wave-cut platform
- Beach profiles and pebble sizes using a transect
- Measuring pebble sizes along a stretch of coast
- Measuring of longshore drift and evidence of longshore drift
- Wave types and frequency
- Questionnaires to visitors
- Field sketches, photographs
- Collecting information on beach protection measures.

SECONDARY DATA

- OS maps
- Geology maps
- Council visitor surveys
- Newspaper reports
- Guide books
- Coastal defences information.

Figure 6.19 Sources of primary and secondary data for coastal studies

To find the height of the cliff:

1 Measure distance X.
2 Measure angle A with a clinometer.
3 Height of cliff = (X × tan A) + height of observer.

3 Indicate on your sketch where the geology and slope angle seem to change and any small scale features, such as joints, caves, landslips, vegetation.

4 Use your secondary sources to help you explain the shape and other features of your cliff profile.

SAFETY

Cliffs and beaches are dangerous.

Take care at all times.

Obey warning notices (Figure 6.21).

Make sure that access to the beach is safe and easy.

Check the times of high tide before starting fieldwork.

Make sure that you will not be cut off by the incoming tide.

Wrap up well against the cold – the UK does not have a Mediterranean climate.

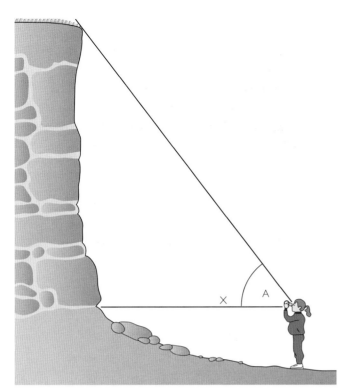

Figure 6.20 Finding the height of a cliff

About sketching cliffs

Look for all or some of the following:

- Width and arrangement of the rock beds.
- Position of bedding planes (lines of weakness between the rock layers).
- Presence of joints or faults (vertical cracks/weaknesses).
- Steepness of the cliff face (vertical, gentle, variations between top and bottom).
- Loose rock at base of cliff/signs of recent cliff collapse.
- The wave-cut notch.
- Related landforms such as caves and arches.

TASKS

Study Figures 6.21a and b.

1 Cliff profiles. Draw profiles for the cliffs in Figure 6.21a in the foreground **(b)** cliffs in the background nearer the point of the headland.

Figure 6.21a Beachy Head

2 Cliff sketch. Make a frame and draw a labelled sketch with the title 'Coastal features at Ramsgate'.

Figure 6.21b Cliffs at Ramsgate

TOP TIPS

- Always observe carefully what the bottom of a cliff looks like.
- This is the action zone.
- Changes in form are occurring all the time.

Beach surveys

When conducting beach surveys it is often necessary to do a series of **transects** in different parts of the beach. They can be either downshore profiles or longshore profiles as shown in Figure 6.22. The downshore profile will show any changes in pebble and particle sizes, and in slope such as between shingle and sand. The longshore profile will show any effects of longshore drift with larger material concentrated at one end of the beach or spit.

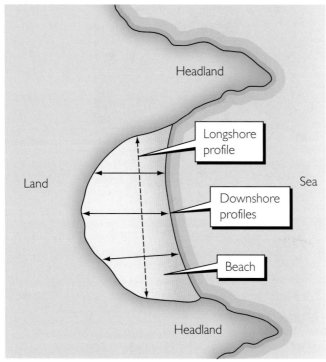

Figure 6.22 Beach transects

About beach profiles

1 Take a general clinometer reading from the edge of the cliff to the low water mark.

2 Record the angle of slope at intervals down the beach. You can do this every metre or select the breaks of slope.

3 Every 2 metres (or less if you prefer) randomly select 5–10 pebbles to measure the long axis and shape (Figure 6.23 and 6.24). You could use a quadrat and random numbers to select your sample as explained on pages 22–23. A good recording sheet, as shown in Figure 6.25, will be needed for recording your results.

Measure the 'a' axis for size

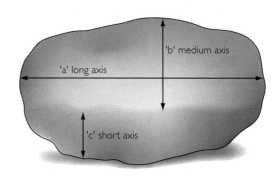

Figure 6.23 How to measure a pebble

Figure 6.24 Measuring pebbles in areas marked out by quadrats

TOP TIP

Consult the tide tables: your beach surveys will be more successful if you do them at low water.

BEACH PROFILE RECORDING SHEET

Location . . . . Weather . . . . Date . . . . Time . . . . .

Site number	Distance from cliff base	Angle of slope	Pebble sizes (cm) (recorded every 2 metres)	Comments
1	1 m	−1°	30, 26, 10, 45, 38 24, 12, 32, 18, 48	No storm beach Large rocks covered in moss/ grasses
2	2 m	0°		
3	3 m	+2°	8, 13, 6, 22, 10 14, 6, 16, 1, 9	Shingle ridge going up

Figure 6.25 Beach profile recording sheet

Tackling Geography Coursework

Other interesting topics centre around investigating the action of the sea:

Destructive waves cause most erosion along our coasts. They are tall, have a strong backwash and much higher wave frequency – there are over 13 waves per minute.
Constructive waves tend to deposit material, are shorter in height, have a strong swash and much lower frequencies – well under 13 per minute.

Measuring wave frequency: this is the number of waves that pass a certain point. To be accurate, count the number of waves over a 5- or 10-minute period and take an average.

Measuring wave height: stand on a pier or groyne and use a metre rule to record the height of the trough and crest of waves as they pass. Average the results of several measurements.

Getting started – good advice

Once you have thought about what you might do and have done some research, the next step is to make a plan.

One example of a plan used by a group of students who were taken to Bournemouth on the south coast of England to do some coursework is given in Figure 6.27. The teacher selected a section of the beach that was popular with visitors and had some coastal protection in the form of groynes. Each student worked as part of a group to collect the data for the first aim. Each student then designed their own extension activities, shown by a * on the plan of one student shown in Figure 6.27.

Title: What are the physical and human features of the beach and cliffs at Bournemouth?

Aims:
1 To describe and explain the physical features of the beach and cliffs
2 To assess the use made of the beach by people
3 To discover the impact of people on the beach

Primary data collection:

Observation:
Photographs, sketches of the main coastal features (aim 1)
Landscape evaluation survey (aim 1)
*Do an environmental quality survey in five different locations (aim 3)

Questionnaire:
Visitors to the beach

Measurement:
At selected sites carry out downshore profiles and pebble surveys (aim 1)
*Groyne survey – measure the difference in the height of the beach material on both sides of the groyne (aim 3)
*Measure footpath erosion on route down to beach (aim 3)

Counts – Record visitor use by taking pedestrian counts at different times (aim 2)

Other data collection
*Write to the local council about the coastal protection techniques (aim 2/3)
Find out about groynes (aim 2/3)

Figure 6.27 Student plan for coastal coursework

When it is completed, check your plan against Figure 6.19 on page 74. Are there any other likely sources of useful data?

Figure 6.26 Part of the beach at Bournemouth. What can be investigated here?

Tackling CBD studies

The CBD is the centre of towns and cities. It is distinguished from the rest of the urban area by having the greatest concentration of shops, banks and building societies, offices, places of entertainment and old buildings. The UK is a highly urbanised country; the majority of people live in towns and cities so that travelling to a CBD to undertake fieldwork is not a hassle for most students.

What can you do?

1 **General topics** (particularly useful if the town/city is not too big)

- What are the main characteristics of the CBD?
- Where are the limits of the CBD?
- Investigation – survey of land uses to show CBD characteristics, or to determine limits, or both.
- What is the sphere of influence of the CBD?

2 Focus on **one theme** (often essential in large towns/cities, for example, with a population over 50,000)

- Where is the core of the CBD in Town/City X?
- Where are the main shopping (retail) areas in the CBD?
- Investigation of shops – classification of types, size, customer counts, customer surveys, shopping quality.
- Hypothesis – shops become smaller, less busy and lower quality moving away from the centre.

3 Examine a **local issue**

- What have been the advantages and disadvantages of pedestrianisation of city centre streets?
- Are the facilities for car parking adequate?
- What changes are taking place in the CBD?
- Investigation of traffic in the town centre – where and why do the greatest traffic problems exist?

Do a survey of street furniture, for example, lighting, hanging baskets, litter bins.

Use questionnaire with shoppers to find the sphere of influence, frequency of visits, goods bought, etc. (see pages 24–27).

Conduct a shopping quality and street appearance survey (see pages 28–29).

Observe what the street is like. This one is partially pedestrianised.

Count the number of storeys for building height.

Survey the shops and services using large scale maps.

Measure the width of shop frontages.

Take shopper counts entering different shops and services.

Figure 6.28 Durham City CBD – what are the fieldwork opportunities?

Tackling Geography Coursework

How can you do it?

The list of data that can be collected is long, but a lot can be collected at one time. It is likely that you will only need to use a selection of the data collection techniques given in Figure 6.29 in your study (say four or five).

Characteristic of the CBD	Data collection technique
1 The CBD contains the major shops and offices, entertainments and public buildings – it is the core of commercial activity.	A land use survey or use a GOAD map. Classify the land used using a system like the one in Figure 6.30. If you are trying to set the CBD limit you need to survey beyond the CBD.
2 The CBD has a wide range of shops and services including convenience and comparison goods.	From your land use survey, classify the shops and services according to convenience and comparison goods.
3 The CBD has the largest shops as determined by shop frontage and building height.	Measure frontage by pacing or trundle wheel, or measure from a large scale OS map. Record the number of storeys of each building as you do your land use survey. You could also record the building use above ground floor level.
4 The CBD has the shops with the highest turnover and largest threshold. A large CBD has large department stores, chain stores and specialist shops.	Threshold populations are available, for example, Sainsbury's needs a threshold population of around 60,000 people. A rough estimate of the thresholds can be obtained by counting the number of customers entering or leaving different shops.
5 The CBD has the largest sphere of influence of a shopping centre in a town.	Use questionnaires with shoppers to find out their home location, frequency of visit, distance travelled, mode of transport, nature of goods bought, etc. You could compare these results for other shopping areas.
6 A core (central area of the CBD) and periphery of the CBD can be recognised.	Your land use survey, building height and frontage survey can be used to recognise the core and periphery. The core is the part with the large department stores, specialist shops, high rise offices and banks; the periphery has smaller shops, cinemas, offices, and on the edge, bus and railway stations, schools and universities.
7 Shopping quality and street appearance decline as you move away from the centre of the CBD.	Use surveys like those shown on page 22 in order to assess the quality of the environment. You could also ask other people's opinions in questionnaires.
8 The CBD contains the greatest volume of traffic and is the most accessible part of a town or city.	Count vehicles in a similar way to pedestrians – make sure you all include the same things, for example, are you going to count bicycles? You could just concentrate on public transport, for example, taxis/buses.
9 The CBD contains the greatest number of pedestrians.	Conduct pedestrian counts in different parts of the CBD. To be most effective you need lots of counts and it is best done all at the same time, for example, as a group activity.
10 The CBD attracts people of all ages and both sexes.	This could be collected through observation of pedestrians or through your questionnaires – don't ask people their sex and take care with age!
11 The CBD is a zone of constant change.	Compare current land uses to old GOAD maps.
12 It has the highest land values in the city including the PLVI (the peak land value intersection).	You need to contact the local Treasury Department or rates office to find this information.

Figure 6.29 Data collection in the CBD

Land use mapping in the CBD

All coursework based in a CBD will need a **land use map** including the types of shops and services. You can collect this data in two ways:

- As **primary** data – complete your own land use survey to show the shops and services using OS maps at the scale of 1:1250 or 1:2500.

- As **secondary** data – from detailed GOAD plans published for most shopping centres in Britain.

The land uses need to be classified. An example of a classification is shown in Figure 6.30a and b. In Figure 6.30b, a recording sheet that could be used is shown.

Field map symbol	Description
A	**Major shopping units**, e.g. department/variety stores
B	**Clothing and shoe shops**
C	**Convenience shops**, e.g. food, tobacconist, newsagent, sweets
D	**Furniture and carpets**
E	**Specialist shops**, e.g. books, sport, jewellers, electrical, hardware, florist, antiques, etc.
F	**Personal services**, e.g. hairdresser, shoe repairs, dry cleaner, launderette, TV rentals, gas/electricity showrooms, travel agents
G	**Catering and entertainment**, e.g. pubs, cafes, hotels, cinema, etc.
H	**Car sales**
J	**Professional services and offices**, e.g. banks, solicitors, architects, doctors, estate agents, opticians, chemists, accountants
K	**Public buildings and offices**, e.g. school, library, Town Hall, Government offices, PO, police station, church, Job Centre
L	**Transport**, e.g. car parks, rail/bus station
M	**Change**, e.g. vacant premises, derelict, under construction
N	**Residential**
P	**Industrial**

Figure 6.30a A simple land use classification for a CBD

Location Silver Street (north side)		Date 24.7.2001		
		Time 4.30pm Weather Fine		
Shop unit size (paces)	**Woolworths** 28	**M & S** 34	**Top Man** 18	**H Samuel** 8
Ground floor use	Variety shop	Food/ clothes shop	Clothing shop	Jewellery shop
1st floor	Offices/ storage	Shop	Shop	Storage
2nd floor	None	None	None	None
Classification	A	A	B	E
Customer count in 10 mins	48	37	17	4

General comments: Pedestrianised street, attractive shop frontages and architecture, hanging baskets, cobbled.

Figure 6.30b Example of a student's data collection sheet. It is for the street shown in Figure 1.03

Getting started – good advice

Where in your chosen CBD?

- This is not a factor if you are doing fieldwork in a small town or city; a complete coverage of the CBD might be possible.

- It is an important consideration if the town or city is large. Either select just one or two locations or do one or two transects by following main routes/roads that pass through the centre of the CBD (see page 22).

When to do the fieldwork?

- Some surveys are best done at a quiet time when there are few workers and shoppers about, for example, a Sunday morning.

  Examples – land use and environmental quality surveys.

- Others are best done at times when there are many people on the streets, e.g. on Saturdays.

  Examples – surveys of shoppers, questionnaires to find out sphere of influence, pedestrian and traffic counts.

- Sometimes you need to visit at different times to bring out variations between peak and off-peak time.

  Examples – traffic surveys and car parking studies.

Example of doing a CBD study

This illustrates, stage by stage, what needs to be done for a complete CBD investigation with the title 'What are the main characteristics of the CBD of …?

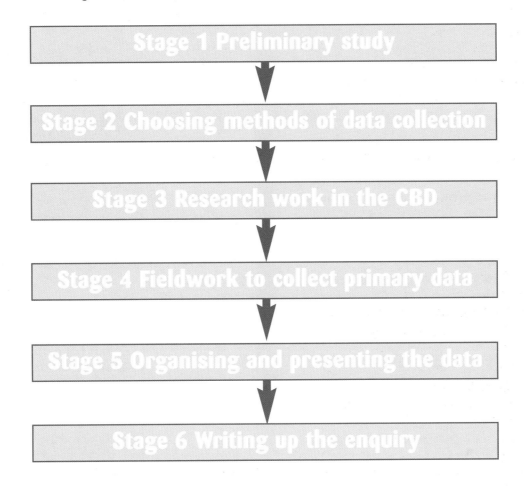

Stage 1 Preliminary study

Stage 2 Choosing methods of data collection

Stage 3 Research work in the CBD

Stage 4 Fieldwork to collect primary data

Stage 5 Organising and presenting the data

Stage 6 Writing up the enquiry

Stage 1 – Preliminary study

Background study – about CBDs:

- Using a geography textbook, make a list of the characteristics of a CBD.
- Focus on those that make it different from other urban zones.
- Decide on three or four characteristics that you would like to investigate.

Thinking – consider what type of fieldwork you prefer to do:

- Stopping people you do not know in the street and asking them questions ☺. Consider studies based on questionnaires such as surveys of shoppers and sphere of influence.
- Stopping unknown people in the street and asking questions ☹. Stick to studies that rely upon observation and counts (as in the following example).

Make decisions – about what to do. Study the following four characteristics:

- **CBD characteristic 1** – land uses are mainly retail (shops) and offices, not housing as in other urban zones.
- **CBD characteristic 2** – buildings are taller in the CBD and floors above street level may also be used commercially.
- **CBD characteristic 3** – large numbers of people are on the streets during the working day, with the largest numbers expected in and near the main shopping area.
- **CBD characteristic 4** – it is the busiest part of the urban area for traffic and for parking.

Stage 2 – Choosing methods of data collection

Method for characteristic 1 (land uses) – map the different land uses in the CBD by observation:

- Make a classification of different urban land uses, e.g. retail (shops), banks and building societies, offices, food and drink (cafes, restaurants and pubs), places of entertainment (cinemas, night clubs, bingo halls), bus stations, parking places, open spaces. Look at Figure 6.30 page 80.
- Make a more detailed classification of retail outlets according to type (e.g. national chain stores, food shops, small shops, charity shops, etc.).
- Make a data collection sheet for recording observations. Look at Figure 6.31 page 80 for ideas.

Method for characteristic 2 (heights of buildings) – count the number of storeys/floors in use and observe their uses.

Method for characteristic 3 (numbers of people) – undertake pedestrian counts at key points within the CBD:

- Make a careful note of locations and the time the counts are made.
- Count for the same length of time at each location.

Method for characteristic 4 – undertake traffic counts at key locations and include counts of parked cars:

- Traffic counts are most likely to be taken towards the edge of the CBD where there are fewer traffic control systems in operation.
- Count parked cars both legal (in car parks) and illegal (on the streets in restricted areas).

Stage 3 – Research work in the CBD

Before you collect data in the CBD:

- Make a preliminary visit.
- Check out the best *places* for carrying out the fieldwork, especially for the pedestrian and traffic counts.
- Assess the amount of work involved.
- Remember – a little time spent on research is likely to save a lot of time doing the fieldwork!

Stage 4 – Fieldwork to collect primary data

When collecting your primary data:

- Choose your times carefully.
- Take note of the previous advice about the best times for different types of survey.

TOP TIPS

How to make people counts more successful

- Can you get other people to help? ☺
- Organise counts at several different locations at the same time.
- Place your helpers at regular intervals from the centre towards the edges of the CBD.
- Make sure that everyone counts at the same time and for the same length of time.
- This ensures a wider and more accurate coverage.

TOP TIPS

What happens when the amount of work is going to be too big?

- Think about taking a 'sample' (pages 22–23).
- Concentrate upon two or three main streets instead of the whole CBD.
- Follow a line of transect along a main road that passes through the centre from one side of the CBD to the other.

Stage 5 – Organising and presenting the data

- Seek to use a variety of techniques of presentation that includes some use of ICT.

- Examples of methods appropriate to the types of data collected in this investigation are:

  - Map of land uses. Devise a colour code for your classification of land uses. Use it on an outline map of the streets of the CBD.

  - Photographs, perhaps taken with a digital camera, to support the land use survey. Remember that it is essential to annotate the photographs by adding labels, in order to highlight land uses shown.

  - Pie graph to show the relative percentages of different land uses after you have added up the total number of units for each type of land use surveyed.

  - A 3D computer graph drawn to represent the number of storeys recorded for the buildings and to show any commercial uses above street level.

  - Bar graphs to show pedestrian and/or traffic counts. If possible, place some or all of them on a map of the CBD, in locations where the counts were made. This has the additional advantage of allowing the pattern/distribution within the CBD to be seen.

  - Isoline map based on people counts. Mark the number counted next to a dot on a map of the CBD. Provided that your counts were taken in at least eight locations, the number of values should be sufficient to allow you to draw isolines. Isolines are like contour lines and are drawn at fixed intervals such as 20, 40, 60 people, etc. This is another way of highlighting where high and low numbers of people were recorded.

  - Pictograph to display the number of car parking spaces available. Different coloured symbols could be used to show any differences between capacity of the car park and number of spaces occupied.

- Remember that the above are only suggestions. You are not forced to use them. You may find other techniques that are equally or better suited to the data you have collected by looking at pages 36–51. Take particular note of the good advice on page 51.

Figure 6.31 CBD in Manchester. What other topic for study is suggested?

Stage 6 – Writing up the enquiry

It is a good idea to divide your coursework into at least the four parts outlined on pages 52–53. You will find that this makes it a lot easier to write up the coursework. There is less chance of missing out information that the moderator will consider important.

Part 1 is the Introduction

Title – What are the main characteristics of the CBD of . . .?

- Main intentions and other aims

- Brief details about the four CBD characteristics to be studied
  - Land uses mainly consist of shops and offices …
  - Buildings are taller and some have two or more storeys …
  - Large numbers of people are present on weekdays …
  - It is the busiest place for traffic and parking …

- General geographical characteristics of the CBD

- Specific information about the town or city being studied (e.g. population, location, access by road and rail, its importance)

- Map of town or city with its CBD clearly identified and labelled on it

Total length 2–3 sides/400–500 words

Part 2 is Data Collection

Details and explanation of methods used:

- Method 1 – Observation of land uses
  - Classification used
  - Example of recording sheet
  - Streets where observations were made
  - Why chosen

- Method 2 – Observation of heights of buildings
  - Example of recording sheet
  - Places/streets where observations were made
  - Why chosen

- Method 3 – People counts
  - Locations where taken
  - Example of recording sheet
  - Time taken and length of counts
  - Explanation for where, how and why

- Method 4 – Counts of traffic and parked cars
  - Example of recording sheet
  - Location, time taken and length for traffic counts
  - Places and times for counts of parked cars

Explanation of how these help with the title and aims.

Summary of tables of data collected.

Tackling Geography Coursework

Part 3 is Analysis and Interpretation

Describe by selecting the key points shown by the data, but do not attempt to describe everything shown in tables and graphs, because most marks are gained by giving reasons. Concentrate upon explanation and keep making comments that relate back to the title and aims.

For example, suggest reasons for land use changes within the CBD and towards its edges. Note any particular land use concentrations and try to offer reasons for them. Explain the results of the pedestrian and traffic counts. Are they related to the pattern of land uses? Look for any similarities and differences between the CBD characteristics of your town or city and the general characteristics of CBDs (that you referred to in the *Introduction*).

Your *Analysis* section is the crucial one for achieving high marks.

Part 4 is the Conclusion

The *Conclusion* is the place to take an overall view of the work. Summarise what your data collection and analysis have shown. You may not have much that is new to say. That doesn't matter; what is more important is that you comment in relation to the title and aims that were stated in the *Introduction*.

The *Conclusion* is easily the shortest part of the writing up. It may only cover one or two sides, but it is very important. Round off the conclusion by relating what you have found for your town or city to that of the general geography of towns and cities. Is your CBD typical? In what ways is it similar to CBDs everywhere? In what ways is your CBD different? Why is it different?

Try to finish in an interesting way. One way of doing this would be to include two photographs of your town or city, one labelled with its typical CBD characteristics and the other with any specific features that make it different from other CBDs (for example, the type of historical buildings).

Studying urban land use zones

Urban land uses include houses, factories, offices, public buildings and transport routes as well as open spaces and areas of unused land. Of these, the one that covers the largest area is housing. In Figure 6.32 a classification scheme is shown, which can be used in urban land uses surveys.

Code number	Code letter
1 Residential	T = Terraced D = Detached F = Flats S = Semi-detached B = Bungalow
2 Industrial	E = Extractive B = Building & construction M= Manufacturing
3 Commercial	S = Shops O = Offices W = Warehouses ø = Other M = Markets B = Banks G = Garages
4 Public buildings	S = Schools H = Local Government O = Other C = Churches E = Other Educational
5 Transport	R = Railways A = Airports B = Bus Stations P = Ports
6 Entertain-ment	H = Hotels C = Cafés L = Leisure Centres I = Inns/Pubs/Clubs T = Theatres/Cinemas A = Arcades
7 Open space	F= Farmland R = Recreation Grounds C = Cemeteries G = Gardens & allotments P = Parks W = Water CP = Car Park
8 Unused land	W= Waste land V = Vacant land being developed D = Derelict buildings

Method

1 Draw a line through the built-up area or follow a road.

2 Record the land use on either side – up to about 100 m if you can. This will give you a belt transect of 200 metres.

3 If you have more time you can do more than one transect going from the town centre to the suburbs in different directions.

4 Use the classification scheme, so if the land use is a terraced house write 1T onto your base map. Here is an example of part of a survey:

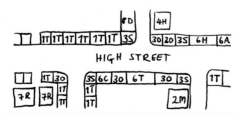

Figure 6.32 An urban land use classification scheme

Only in small settlements can one person or a group hope to study the whole area. The only practical method in urban areas is to select one or more transects across the built-up area (see page 23). One of the most popular themes is to investigate the differences between urban zones and the urban models.

Using urban models

There are several different models, but the most widely used are those of Burgess and Hoyt. Most of the urban models refer to:

- Functional zones – the location of shops, services, housing and industry.

- Ages of buildings – decreasing in age towards the outskirts. Figures 6.34 to 6.36 show the characteristics of housing of different ages and how their locations are different in large cities such as Manchester.

- Social class of areas – higher class properties are usually towards the outskirts; in UK cities they are more likely to be on the western side of the CBD, away from the air pollution that was carried by the westerly winds to the properties in the east after the Industrial Revolution.

- Quality of the housing (including size and amenities) and the environment – increasing towards the outskirts.

TASKS

1 (a) Draw a diagram to show one model of urban land use.

   (b) Add labels for the likely locations of the three housing types in Figures 6.34 to 6.36.

   (c) Describe how Figures 6.34 to 6.36 suggest that the quality of housing increases towards the edge of a city.

2 From Figure 6.33 state the evidence that the CBD of Manchester lies north of grid line 97.

3 The land use transect used by the student began just north of Oxford Road railway station (840973) and followed the B class road (B5093) to East Didsbury railway station (853903).

   (a) Explain why this was a good choice of route for an urban land use transect.

   (b) Why would a transect route further west along the A5103 not have been as good?

4 In a city the size of Manchester, why is it:

   (a) possible to survey only one transect route

   (b) necessary to take a sample of land uses instead of making a full survey?

INFORMATION

- For transects and sampling – look at pages 22–23.
- For questionnaires – look at pages 24–27.
- For environmental quality surveys – look at pages 28–29.

TOP TIPS

- Always seek help from your teacher in the planning stage.
- Discuss with your teacher the number of transects to be surveyed.
- The teacher knows best about what is possible and what is needed.

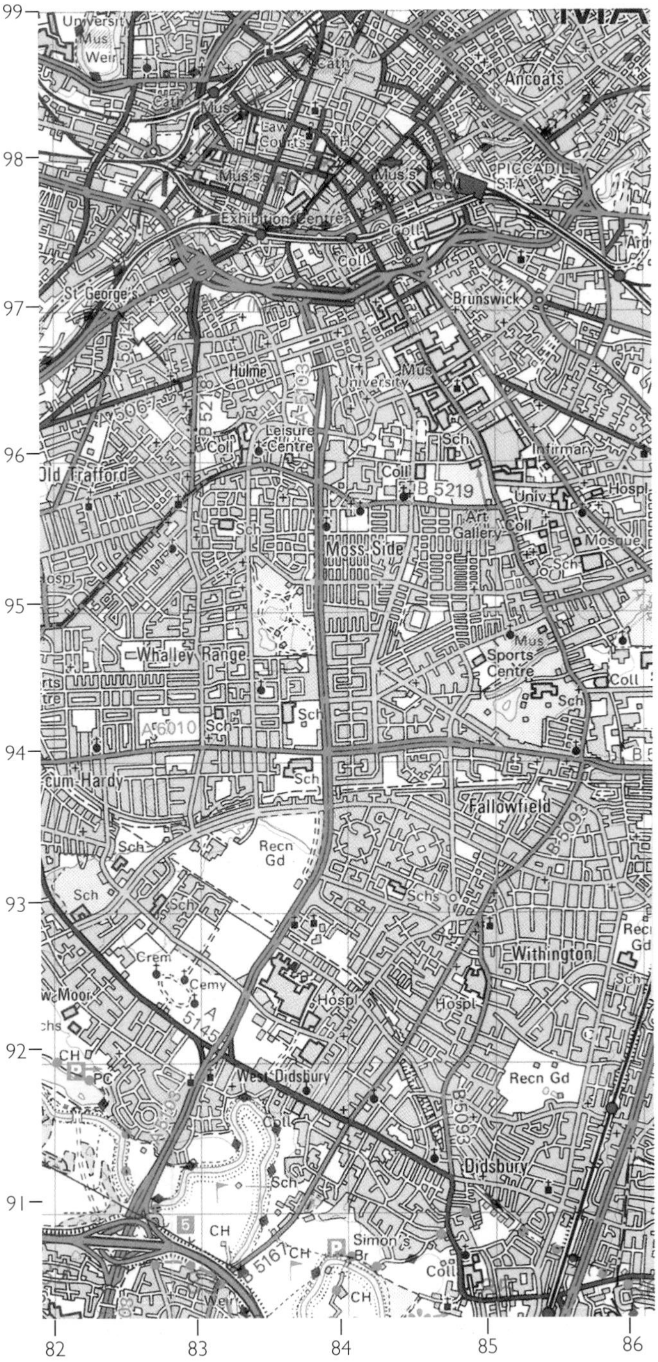

Figure 6.34 Victorian terraced housing

Figure 6.35 Interwar/postwar semi-detached housing

Figure 6.36 Modern detached housing

Figure 6.33 Manchester 1:50 000 OS Map

Leisure and tourism

Leisure and tourism are popular topics for coursework and they may be studied in either urban or rural areas. However, don't fall into the trap of producing a project full of pretty brochures, a tour guide of your local city or a list of activities and prices in the local leisure centre! Such projects will not gain you many marks. In fact many can only be described as disaster areas!

Most leisure and tourism studies include at least one of the following themes:

- Distribution of leisure and/or tourist facilities
- Use made of leisure and/or tourist facilities
- Impact of leisure/tourism on an area and its people, for example, at a honeypot site in a National Park (Figure 6.37)
- Methods of management to limit damage caused by visitor pressure
- Sample of people's use of leisure time for recreation or holidays.

Leisure

The table below gives you some ideas about what to study for recreation and leisure and how to collect the data. Beware of only studying one leisure centre – it often turns into a catalogue of activities and prices, which is not what is required.

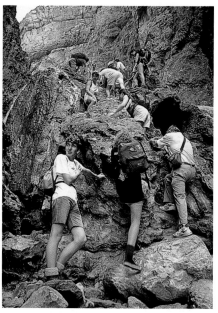

Figure 6.37 Gordale near Malham in the Yorkshire Dales National Park – a classic example of a honeypot

Inquiry idea	Fieldwork	Useful secondary sources
The distribution and use of recreational facilities: – consider targeting one section of the population, e.g. young people or older people – select a small range of facilities, e.g. sports facilities, open spaces and parks in one part of a city.	Choose a well-defined area – either a few grid squares in an urban area or a rural district council area. Map the location of leisure facilities. Use questionnaires to find out the use and spheres of influence (or catchment areas) of different facilities.	Visit Local Authority/Council websites.
A comparison of the spheres of influence of different leisure facilities, for example, a public house, tennis club, sports centre and theatre. You could try to establish a hierarchy of facilities.	Use questionnaires at the locations or as a house-to-house survey. Map the facilities. Do counts of visitors to the facilities at different times of the day and week and/or car park surveys.	Membership lists from clubs and societies, visitor figures from local tourist boards and councils.
A comparison of two or more parks – either in an urban area or country parks.	Map the area, recording the number of types of facilities available. Record the size and general location. Use photographs and sketches. Take visitor counts at entrance(s). Measure spheres of influence by questionnaire, also frequency of visit, reason for visit, mode of travel. Do environmental quality and litter surveys. Interview park wardens about management.	Council information leaflets about leisure provision, visitor numbers and local transport information to show accessibility.
Sample a section of the population and investigate their leisure patterns – over a day, a week or for holidays. This can be used to compare leisure patterns of different age groups and/or income groups.	Questionnaires will need to be carefully worded; also take care in choosing your sample. Map the locations of facilities and measure distances travelled by the users, etc.	National or local statistics on leisure, so that your results can be compared with the general pattern.

Example of planning an urban leisure enquiry

1 Which leisure pursuit to study?

List the possibilities in your town:

- Cinema
- Bingo hall
- Swimming pool
- Football ground
- Pub
- Park
- Leisure centre.

Make a decision:

- The main park.

Justify your decision:

- There is more to see and do in the park.
- There are landscape features that can be described. This is what geography is all about.
- There won't be great hordes of people arriving and leaving at the same time.
  People will be coming and going all the time.
  This will make it easier to give out questionnaires.
- The park is open to the public – access will be easy. You have to pay to go into the others.

2 Find a title

List some of the possibilities for study in an urban park:

- Relationship between the type and distribution of facilities and the physical geography of the park – relief (height and slope angles), drainage (surface streams and lakes) and vegetation (trees and other plants).
- Attractions and facilities in the park and whether they are suitable for all age groups.
- Do the visitors consider the type and number of facilities provided to be adequate?
- Could the range of facilities in the park be improved to meet people's needs? If yes, how could they be improved?
- Investigation of the amount of use made of the different facilities to suggest how the park could be better managed for visitors.
- What is the sphere of influence for people visiting the park?
- What are the views of visitors and local residents about the park?

Make a decision:

- I have decided to investigate the sphere of influence of the park.

Justify your decision:

- I like meeting and talking to people.
- I know that asking questions and handing out questionnaires is a good way of getting a lot of information fast.
- It will still be relevant to observe and describe what there is in the park.
- I wouldn't be happy measuring slope angles and doing vegetation surveys.

3 Data collection

Methods of data collection:

- Main method – do a questionnaire and expect to get at least 50 filled in.
- Describe the physical landscape with photographs and sketches
- Map the facilities provided for people.
- Look to see if there is any information about the park on the Council's website; if not, contact the Council's Parks and Leisure offices to find out if they have any information on who visits the park.

TASKS

1 Make a questionnaire suitable for an investigation with the title – 'How large is the sphere of influence of the urban park in City X?'

2 The questionnaire is the main method of data collection. Explain what can be done to try to make sure that enough data is collected.

Tourism

Studying the impact of tourism at a honeypot site is a popular choice for coursework. A honeypot is a particularly popular tourist location. Examples include historical sites such as Stonehenge, certain areas in National Parks where there is a concentration of natural attractions and major tourist towns such as York and Oxford. When making a study of the **impact** of tourism, you must look at both the advantages and disadvantages that visitors bring to an area.

An example of urban tourism

A group of four students brainstormed the idea of doing some coursework about the impact of tourism in York. York is the most visited tourist city in the north of England (Figures 6.38 and 6.39). Figure 6.40 shows the results of their brainstorming session. From these ideas, students planned their own data collection.

Figure 6.38 Two major York attractions – the Minster and its well-preserved Roman walls

Figure 6.39 Sign board on the main tourist trail in York

Map the tourist attractions.

Traffic flow and car park surveys in the tourist season and in the off-peak time.

Investigate the impact on the local economy by interviewing local shop owners and people who work in the tourism industry.

Environmental quality/litter surveys in tourist areas at different times of the year.

Count visitors.

Collect tourist information and literature on hotels and guest houses and occupancy through the year.

Assess the quality of the tourist attractions to see if the facilities that score highest attract the most visitors.

Survey the shops and services and note which ones would be mainly use by tourists; interview local people to see if any conflicts exist.

Use the Census Small Area Statistics to find out how many people are employed in tourism.

Figure 6.40 Student brainstorm 'To investigate the impact of tourism in York in summer and winter'

Another student drew up a visitor questionnaire to find out basic information about visitors to York and to discover their likes and dislikes. This is shown in Figure 6.41. What are the strengths of this questionnaire? How could it have been improved?

Tackling Geography Coursework

VISITOR QUESTIONNAIRE

I	In which town (in the UK) or country (foreign visitor) do you live?
2	How have you travelled here (from within the UK)? Car Bus Train Cycle Other
3	How long are you staying? Less than I day 2–3 days 5–7 days I–2 weeks Longer
4	In what type of accommodation are you staying? Hotel Guest house/B&B Friends/Relatives Camping/caravan Others
5	Rate the following on a scale of I–5 (I = poor, 3 = average, 5 = good)

	1	2	3	4	5
York Minster					
Jarvik Museum					
The Walls					
The Shambles					
The Castle Museum					
Shopping					
Eating					
Parking					
Entertainment					
Service by the local people					
Tidiness					

6	What do you most like/enjoy about York?
7	What do you dislike/not enjoy about York?
8	Have you visited the Tourist Information Office? Yes No

Figure 6.41 A student's questionnaire for visitors to York

An example of rural tourism

Figure 6.42 shows another student's ideas for data collection in a Forestry Commission plantation, in which tourism is encouraged. Facilities and services have been provided for visitors. Study the student's plan for data collection. Can you recognise the different methods of data collection that will be used?

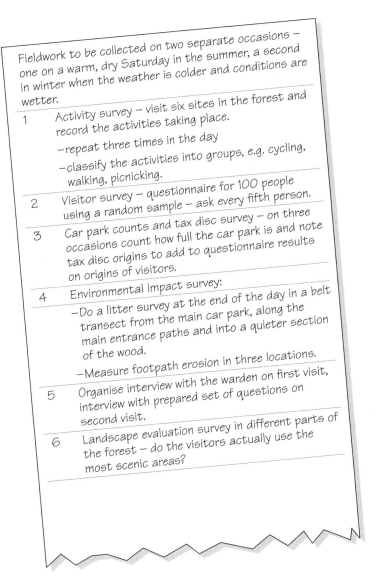

Fieldwork to be collected on two separate occasions – one on a warm, dry Saturday in the summer, a second in winter when the weather is colder and conditions are wetter.

1   Activity survey – visit six sites in the forest and record the activities taking place.
    – repeat three times in the day
    – classify the activities into groups, e.g. cycling, walking, picnicking.

2   Visitor survey – questionnaire for 100 people using a random sample – ask every fifth person.

3   Car park counts and tax disc survey – on three occasions count how full the car park is and note tax disc origins to add to questionnaire results on origins of visitors.

4   Environmental impact survey:
    – Do a litter survey at the end of the day in a belt transect from the main car park, along the main entrance paths and into a quieter section of the wood.
    – Measure footpath erosion in three locations.

5   Organise interview with the warden on first visit, interview with prepared set of questions on second visit.

6   Landscape evaluation survey in different parts of the forest – do the visitors actually use the most scenic areas?

Figure 6.42 Student plan

TASKS

1   What are the advantages of having a student brainstorming session before starting coursework?

2   **(a)** List the good and bad points of the student's visitor questionnaire in Figure 6.41.

   (b) Make another version of the questionnaire with improved questions and layout.

3   Study the student plan in Figure 6.42. Of the six items:

   (a) Which ones depend on observation?

   (b) How many rely upon asking people questions?

   (c) Which other technique of data collection will be used?

Vegetation and soils

Vegetation and soils are good geographical topics for coursework but you do need to plan the data collection carefully. The time of year is particularly important for vegetation studies (Figure 6.43).

Figure 6.44 shows different types of data collection for topics related to vegetation and soils. Many permutations for study are possible. You can concentrate on vegetation only, on vegetation and land uses, on soils only, on vegetation, soils and land uses or upon human effects of the rural landscape that can be observed in places that receive many visitors (such as along footpaths in National Parks).

Figure 6.43 Woodland in summer, the season with the best opportunities for fieldwork. What possibilities exist?

PRIMARY DATA COLLECTION

- Recording vegetation and land uses along transect lines
- Surveying plant types and % cover using quadrats
- Collecting soil samples using a trowel or soil auger to analyse soil characteristics
- Measuring slope angles and relating them to changes in vegetation, soils and land uses
- Observing the natural environment using landscape evaluation, photographs and field sketches
- Taking transects across footpaths, measuring width, depth and % vegetation cover, to ascertain amount and effects of footpath erosion caused by humans.

SECONDARY DATA COLLECTION

- Geology books and maps
- Guide books for plant identification
- Text books about soil studies.

Figure 6.44 Possible data collection for vegetation and soil studies

Use a recording sheet like the one in Figure 6.45 for your results.

Site number: 2   Grid reference: Height: 230m
Aspect: N   Slope angle: 15°   Geology: Granite
Soil depth: 42cm   Land use: Rough pasture/fell
Site observations: Exposed hill slope on valley side, ground quite dry

Vegetation species	Number plants present	Estimated % cover	
1 Heather	ɪɪɪ ɪɪɪ ɪɪɪ = 13	63%	
2 Bracken	ɪɪɪ ɪ = 6	25%	
3			
4			
5			
6 Bare ground	ɪɪɪɪ = 4	14%	

Bracken

Heather

Bare ground

1m

1m

Each small square = 1%

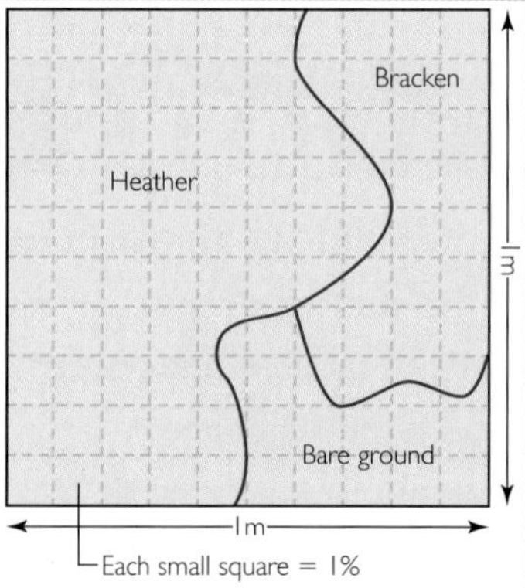

Figure 6.45 Recording sheet for a vegetation transect

Studying vegetation

How vegetation changes across a landscape is best done by using sampling along a transect (see page 22). You need to choose the location of your transect carefully. Try to include as much variety as possible, for example, areas with a different slope, rock type and altitude, so that a range of vegetation types exist. The intervals used for sampling along the transect mainly depend upon the vegetation – areas with large trees do not need to be sampled as often as a grassland area. It is often a good idea to use a belt transect, say 1–2 metres wide (or wider in an area with large trees).

Use of quadrats is common in vegetation surveys. Quadrats are usually 0.5 or 1 metre square. Their use in a footpath erosion survey is shown in Figure 3.22 on page 35. Within each quadrat you can measure the:

- Number of different species
- Percentage cover of each plant species
- Frequency of each plant species (e.g. if a plant was present in 10 out of 20 quadrats then the frequency would be 50%).

Studying soils

Soil study is only easy in places where the soil profile is exposed – such as on an exposed river bank or the sides of a cutting on the road. Without digging a pit (and upsetting the farmer!) the technique for collecting a soil sample is by using a soil auger (Figure 6.46). This can be used to find the depth of soil and different horizons or layers. Each horizon in a soil often has a different colour, texture and pH. Some basic soil tests you can do in the field are shown (right).

Figure 6.46 Soil auger

Human impact on vegetation: footpath erosion

Footpath erosion and the trampling of vegetation can be measured by doing a transect 10 metres either side of a footpath or track. You can measure the depth and width of erosion, as well as percentage cover and identify differences in the types of vegetation.

SOIL TESTS

Colour

Example of soil colours in a profile:

Top of the profile

↑

■ Good drainage, high organic content

■ Minerals washed out of the soil

■ Iron rich mineral deposition

□ B horizon

↓

Bottom of the profile

Figure 6.47 Example of soil colours in a profile

Texture

1 Moisten the soil.
2 Rub between fingers to feel for stoniness.
3 Will it roll into a ball easily?
    | \
  No   Yes—SAND Gritty, not
  | sticky
4 Will it roll into a thin thread, less than 5 mm?
    | \
  No   Yes—LOAM not sticky
  |
5 Will it roll into any shape without cracking, e.g. a ring or horseshoe?
         \
      Yes—CLAY sticky

pH

The measure of soil acidity on the scale of 1–14. Low numbers are acid, a pH of 7 is neutral and high numbers are alkaline. pH affects vegetation, land use and soil forming processes. Soil pH can be measured using a pH meter or a pH testing kit.

Investigating weather

In order to satisfy the Examination Boards' requirement that a complete coursework submission must include primary data, you will need to set up weather instruments in one or more locations, and take readings over a long enough period to make the measurements useful. Having collected primary data, there are plentiful opportunities to use secondary data in order to compare your measurements either with previous records for your own location or with records from other locations taken at same time.

PRIMARY DATA

- Measurements using weather instruments of wind speed and direction, precipitation, temperature, humidity, pressure and sunshine
- Observations of amount of cloud cover, types of cloud and present weather.

SECONDARY DATA

- Data downloaded from websites and automatic weather recording stations
- Weather maps, reports and daily data from newspapers
- Weather reports from TV and radio
- Information and reports from your local weather centre/met office.

Figure 6.48 Sources of data for weather investigations

Urban and rural weather studies

The majority of you live in urban areas where there are plentiful opportunities for comparing weather readings in different locations within your town or city.

Example titles include:

- Hypothesis – Temperatures decrease with increasing distance from the city centre.
- Does an urban heat island exist in the centre of city X?
- Is there a relationship between temperature and building density?
- What changes in temperature and humidity occur along a transect from city centre to city edge?
- Are there differences in temperature inside the large city park?
- Do large buildings affect the speed and direction of the wind?

If you live in a rural area, the main opportunities for local study are offered by variations in relief. Example titles include:

- How and why do weather conditions change across the valley?
- How large are the differences in temperature, precipitation and wind speed between the lowland and upland parts of Farm X?

Taking measurements

To measure the weather yourself, you either need to set up your own instruments or use a school weather station. Figure 6.49 shows many of the instruments used in a school weather station. Figures 6.50 and 6.51 illustrate ways to approach two different weather studies, one urban and one rural.

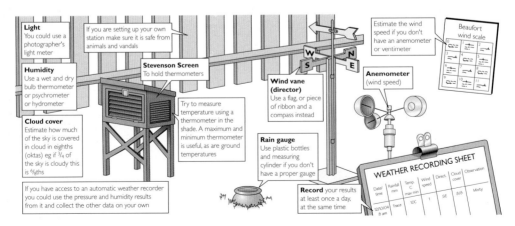

Figure 6.49 Collecting primary data – methods of measurement in a school weather station

Examples of weather studies

1 Aim – Investigation of variations in weather readings in a large garden (or park). Look at Figure 6.50.

Method:

- Decide upon sites. The more the better!
- Do regular, accurate recordings at set times.
- Take notes, photos, sections about each site.

2 Aim – To investigate the microclimate across a valley. Look at Figure 6.51.

Method:

- Select the transect to incorporate as much variety as possible. A North–South transect is best.
- Select sites to record data – ten is usually enough.
- Decide timing of data collection.
- Collect data at about the same times for several days.
- Observe basic information about each of your sites, for example, vegetation, land use, slope angle.
- Design your data recording sheets to take both information about each site and the weather measurements.

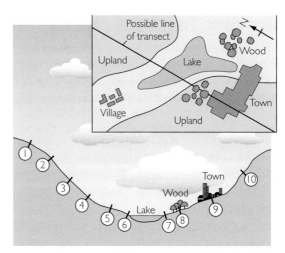

Figure 6.51 Investigating microclimates in a valley

In Figure 6.51, sites 1–5 can be used to investigate the impact of slope angle, time of day, aspect and altitude. Sites 6–9 were used to investigate the effects of the lake, wood and town.

- Remember: you will need ten rain gauges if you collect rainfall at each site and you will need to move quickly between sites if the data is going to be reliable.

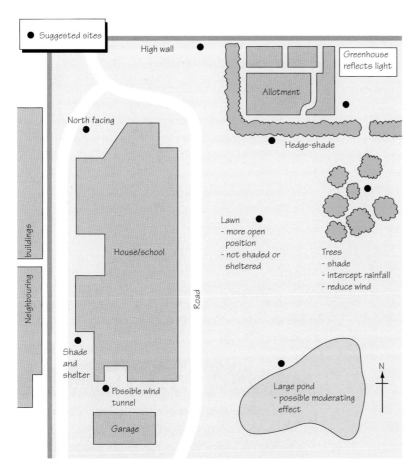

Figure 6.50 Investigating microclimates in a large garden

Figure 6.52 An important source of supplementary secondary weather data

Coursework in rural areas

Many physical geography studies take place in rural areas – such as those tackling rivers or investigating vegetation and soils. This short section concentrates upon coursework that is predominantly human. Access to study areas can be more of a problem in the countryside than in urban areas, since most land is in private ownership. Unless you can gain permission first from the landowner, your choice of study sites will be restricted to those along public footpaths and where there are public rights of access.

Example titles

Land use surveys

- How and why are land uses different between Area X and Area Y?
- In what ways and why do land uses change between the valley floor and hilltops?
- Hypothesis – rural land uses become more varied as you travel down valley/from upland to lowland areas.
- What are the similarities and differences between present and past land uses in Area X?

Investigating farms

- What physical and human reasons explain variations in land uses between Farm X and Farm Y?
- In what way and why have land uses on Farm X changed in the last five years?
- Have farmers in Area X diversified into activities other than farming?

Rural activities other than farming

- What are the impacts of Quarry X on the environment and people of the surrounding area?
- Should the company be given permission to increase quarry size?
- What are the advantages and disadvantages of tourism in Rural Area I?
- Is Village X a tourist honeypot?

PRIMARY DATA

Observing:
- Land use surveys
- Landscape evaluation surveys supported by photographs and field sketches.

Measuring and counting:
- Physical characteristics such as slope angles, vegetation coverage and soil types
- Traffic and people counts.

Asking questions:
- Interviews with farmers
- Questionnaires for both local people and visitors.

SECONDARY DATA

- Large scale maps, e.g. OS, land use and geological maps
- Government statistics about farming and rural affairs from www.defra.gov.uk
- Newspaper articles about local issues, for example, tourists and quarrying.

Figure 6.53 Methods of data collection for studies in rural areas

Land use surveys

The key to making a success of a study of land uses in different areas or on different farms is to choose two or more examples with a plenty of variation between them, otherwise your opportunities for explanation and comment will be very limited when it comes to writing up the work.

If you are attempting to relate land uses to physical conditions, the best advice is to follow a transect line. Choose one that crosses an area with marked changes in land use and physical conditions. Good examples of this are from the valley floor on to an escarpment, or from one side to the other across a wide river valley with steep sides. Make your transect quite wide, say about 8 kilometres, to allow you to follow roads and tracks across the area, from which land uses on either side can be recorded in order to complete the transect.

TOP TIP

Don't leave it until winter to do farm and rural land use surveys – the crops have gone, some animals will be indoors and all the fields may look the same (especially if snow has fallen!).

Class	Altitude (m)	Wetness No...	Soil quality Deep soil	Soil fertility (pH)	Slope (°)
1 High quality	Below 100	Free drainage Rainfall <750mm	75cm+ Stone free Loam texture	7+ (neutral)	Level (not above 3)
2	100–150	Imperfectly drained Drainage easily modified by liming	Depth 50–75cm Slightly stony	6.0–6.5	Slight (not above 7)
3	150–200	Some problems but possible to in stall drainage system texture	Depth 25–50cm Stony – may be sandy or clayey	5.5–6.0	Moderate (not above 11)
4	200–350	Poorly drained but can be improved to maintain pasture	Shallow – under 25cm Very stony	5.0	Significant (11–20)
5 Low quality	Above 350	Poorly drained Drainage almost impossible to install. Rainfall >1250	No humus Very stony – skeletal soil only	Under 4.5	Steep (over 20)

Figure 6.54 Relationships between land quality and physical conditions

Investigating farming

The most successful farm studies tend to use more than one farm, because you can compare farming systems, land uses and changes. Studies are especially successful if your chosen farms are in different physical regions and specialise in different types of farming.

In your farming study you can investigate some or all of the following:

- Land uses on the farm and how they are linked to physical and economic factors
- The farm as a system with its inputs, processes and outputs
- Patterns of work at different times of the year
- How the land use pattern and farm system have changed (provided that you can have access to old farm records) and are continuing to change
- Diversification – farmers receiving income from non-farming activities such as running farm shops, campsites, rural crafts, etc.

Interviewing farmers

Without the farmer's cooperation this type of investigation is impossible. As a rule some kind of personal introduction is needed. Even then, you must remember that farmers are busy people, particularly at certain times of the year. Follow these guidelines with your interviews:

- Avoid hay and harvest times.
- Make an appointment in advance.
- Be well prepared and ready with your questions.

- Don't have too many questions – a maximum of 15 questions is a good guide.

Quarrying

Quarries cause a lot of controversy in rural areas. There are often conflicts of interest between the quarry owners (who think mainly in terms of profit), local people (who need the work) and tourists (who complain about landscape damage and pollution). There are good opportunities to use questionnaires to local residents, measure noise and dust pollution, and to consider the economic uses of the rock being quarried.

Other issues in rural areas

Any proposed new developments through or in rural areas, such as motorways, by-passes, out-of-town shopping centres, business parks, housing estates and golf courses are bound to have both supporters and objectors. This is why rural issues like these can be very good subjects for investigation. It is possible to question people on both sides of the conflict and weigh up the relative strengths of the opposing views. Useful secondary sources of information are articles in the local newspaper websites set up by interested parties to win support for their views.

Village studies

If you live in or near a village there are four likely themes for geographical study:

- Change (this may be growth or decline)
- Local issues (e.g. does the village need a by-pass?)
- Problem solving (e.g. are there sufficient services for the local inhabitants?)
- Comparisons between it and one or two other villages.

Figure 6.54 shows you the types of data you can collect as part of a village study.

Investigating changes

1 Land use changes

Old maps, from a variety of different dates, will show land use changes and how the village has grown. Your own land use survey will be more up to date than any published map. Identify:

- The main housing periods, for example, pre-Victorian (1830s), Victorian/Edwardian (1830–1918), inter-war (1918–39), post-war (1945–70) and modern/recent (see pages 86–87)
- The changes in shops and services.

2 Population changes

Parish records, sometimes held in the records office or reference library in the nearest large town, are the main source for earlier times. Study the information with care, because sometimes figures for parishes include more than one village and several surrounding farms.

Figure 6.54 What could I study in my village?

Questionnaires are the best method for discovering details about today's inhabitants – sex, age ranges, family sizes and how long they have lived in the village (social information), occupations and place of work (economic information). There may be opportunities to investigate different population characteristics between one part of the village and another, especially if there is a modern housing estate with recent arrivals, which would allow old and new parts of the village to be compared.

3 Changes in function, shops and services

In earlier centuries farming was the dominant economic activity in villages. Today the majority of village inhabitants travel elsewhere to work. Their function has changed from a farming settlement to a commuter village. Growth in leisure and tourism is another change, especially in villages located in or near a National Park. There may be an opportunity to investigate the good and bad impacts of tourism on the village environment and people.

Large villages may still have a range of services (post office, grocers, newsagents, garage, pub, hairdressers). One possibility is to find their sphere of influence using questionnaires. Ask your questions outside the different services. Count numbers of customers and numbers of vehicles passing through the village that stop to use the services. You may be able to make a judgement about whether the shops and services are likely to survive.

PRIMARY DATA

- Land use survey
- Survey of shops and services
- Environmental quality survey
- Questionnaires about the people, their habits and opinions
- Traffic counts.

SECONDARY DATA

- Old photographs and maps
- Parish records
- Newspaper articles and books about the village
- Census data
- Estate agents for types of house and prices.

Figure 6.55 Data collection for village studies

TOP TIPS

- Don't just call your enquiry 'A Village Study' – it is very important that you have a theme.
- Concentrate on the present – don't overload it with historical information (see page 54 for the dangers).
- Get the balance right between primary and secondary data – you need more primary than secondary.

Investigating issues and problems

Some villages are in danger of losing their remaining shops and services. This could offer a fertile area for investigation. Map the shops and services that survive. Use well thought out questionnaires and ask residents which ones they use, how often, why and why not. You could broaden out the questions to discover their shopping patterns in general. Think carefully about selecting your sample of people to be questioned – will older, retired, non-car drivers use the village services differently to a younger car-owning family with children? Look back to pages 22–23 for more information about sampling.

As a result of the great increases in road traffic, one issue for many villages is whether a by-pass should be built. Figure 6.56 shows the plan that was used by one student before the by-pass at Witton Gilbert, a dormitory village in County Durham, was opened. The coursework investigated whether Witton Gilbert needed a bypass. The by-pass has now been completed; it is in the foreground on Figure 6.57. Another student wishing to investigate the impact of the by-pass could compare their own results with those of this student.

Title: Does my local village need a by-pass?

Data collection plan

Observation

Take photographs and field sketches
Landscape evaluations to test which, if any, would be the best site for the new road.

Questionnaires

Plan and collect questionnaire responses. I aim to collect 60 questionnaires from local residents – 30 who live in the old part of the village and 30 who live in the new housing estates. A further 40 questionnaires from the local school, the developer, shopkeepers in the village and the garage. I aim to sample a cross-section of older and younger residents.

Interview the local council representative.

Measurement

Do traffic counts at different times of the day for one week.

Others

Visit library to collect information on the development and changes in the village.
Study maps to decide where the by-pass might go.

Figure 6.56 Student plan before the by-pass was built

Comparing two or more villages

Doing this is beneficial when:

- Villages in your local area are small – investigating one will not yield sufficient data and allow adequate comment for a top grade
- One village was studied in a teacher-led enquiry – your individual investigation of a second village will extend the enquiry and give you access to the marks allocated for individual initiative (see page 6).

Figure 6.58 shows you the title page and data collection headings from a student's coursework that compared two villages. The same data were collected in both villages.

Figure 6.57 Witton Gilbert with by-pass completed. Was everyone in the village pleased?

Title: What are the similarities and differences between Brandon Village and Quaking Houses?

Land use surveys

Data collection: Secondary source information – (a) village development, (b) population growth

Surveys of shops and services

Questionnaires on population and employment

Environmental quality surveys

Figure 6.58 Comparing two villages

Environmental pollution

Pollution of air, land and water are problems in both urban and rural areas. Figure 6.59 summarises how you could collect data in a pollution survey. Official organisations that measure pollution have costly equipment to survey pollution, but there are some simpler methods to use – as you will see on these pages.

PRIMARY DATA

- Litter, noise, air and water pollution surveys
- Landscape evaluation survey (see pages 28–29)
- Discover people's perceptions of the problem by using questionnaires
- Observe supported by photographs and sketches.

SECONDARY DATA

- Environmental agencies and local councils
- Water Boards
- Environmental or other groups campaigning on local issues
- Newspaper articles and reports.

Figure 6.59 Data collection techniques for pollution surveys

When can you use pollution surveys?

1 A pollution survey could form the main theme for the coursework, for example, looking at the variations in pollution in an urban area or along the course of a river.

2 Alternatively, a pollution survey could form part of the data collection in a wider investigation, such as a river study or beach survey, or the impacts of economic activities such as quarrying or tourism and its consequences.

Visual pollution

This is a measure of land pollution and you can use a scale like the one shown in Figure 6.60. Add up the scores for each area; the higher the score the more polluted the area. Litter counts could form the basis for a separate investigation. Remember the scores are your own personal perception – someone else may have given a different score.

Noise pollution

Check at school – the science or geography department may have a sound meter that would be more accurate. If not, use a descriptive scale for noise levels, such as 1 = no apparent noise at all through to 5 = so loud that you cannot hear yourself speak. Study noise levels at different times of the day or the week and with increasing distance from the source such as a road, airport, quarry or factory.

Aspects of pollution	Score No pollution ◄──► Badly Polluted				Guidelines
	0	1	2	3	
Size of site					Larger the size, worse the pollution
How obvious to passers-by					The more obvious the pollution, the higher the score
Colour and texture					Heaps of rubbish, worse than a thin cover
Impact on surrounding area					If in a pleasant area the pollution scores higher than if in a poor area
State of buildings					Old, derelict buildings nearby score high
Abandoned cars, etc.					Many large items such as old beds, fridges, etc., score high
Litter					Greater the density, higher the score
Smells					Worse the smell, higher the score
Vegetation cover					Plants can hide pollution, so the greater the vegetation, the lower the score

Figure 6.60 Scale of visual pollution

Water pollution

You need to choose your river or stream carefully – as a rough guide, streams with stagnant stretches may be the best sites to find evidence of pollution.

State of water	Animals present
Clean, unpolluted water	Stonefly larvae, Mayfly larvae, Salmon, Trout, Grayling
Doubtful	Caddis-fly larvae, Freshwater shrimp
Poor	Water louse, Blood worm (Midge larvae), Leech, Roach, Gudgeon
Grossly polluted	Sludge worm, Rat-tailed maggot

Figure 6.61 State of water and animals present

Method

1 Survey the area of study and note the possible pollution sources and select about ten sites.

2 Complete a recording sheet like the one in Figure 6.63 for each of your sites.

3 Carry out a survey of the water life between spring and summer. Figure 6.62 shows some indicator species to look for. Try to count the number of different species at each site – usually there will be about 20 different species in non-polluted water.

4 Measure water temperature – hot water discharges are a pollutant and they affect the amount of dissolved oxygen in the water.

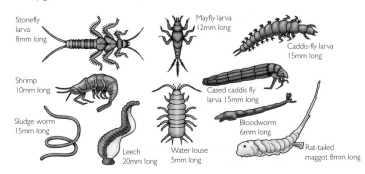

Figure 6.62 Biotic index

River pollution booking sheet		
River, stream:	Site number:	Grid ref.:
Date:	Weather:	Depth of river (cm):
Width (m):	Ave. surface velocity (m/sec):	Visibility using yoghurt carton (cm):
Water temperature (°C):	pH:	

VISUAL SURVEY (✔ tick)	0 Points	1 Point	2 Points	3 Points	4 Points
Presence of suspended solids, e.g. sewage	V. clear	Clear	Fairly clear	Slightly murky	Murky
Colour	V. clear	Clear	Slightly brown	Dark brown	Black
Stones	Clean and bare	Clean	Lightly covered in brown fluffy matter	Coated with brown fluff	Coated with brown and grey deposits
Water weed	None	A little in shallows	Lots in shallows	Abundant	Chocked
Grey algae (sewage fungus)	None	None	A little	Present in patches	Plentiful
Scum/froth/oil	None	Odd bubbles	Noticeable foam islands	Large quantities	Covers whole river
Dumped rubbish	None	A few small items	A few large items	Large and small items	Many large, different items

Overall score: 0–3 very clean; 4–9 clean; 10–15 fairly clean; 16–21 doubtful; 22+ badly polluted

Figure 6.63 Recording sheet for river pollution

Investigating settlement and shopping hierarchies and transport

Settlements and shopping centres can be arranged in order of size and importance. This is called a hierarchy. You can arrange either settlements or shopping centres into order according to:

- Size
- Sphere of influence (trade area)
- Goods and services available.

Settlement hierarchies

To investigate a settlement hierarchy you would need to select a variety of settlements of different size in your local area. For each settlement:

- Find out its population size using census or other secondary information locally available.
- Do a survey of the shops and services (see page 86). You need the total number in each settlement as well as the percentages of high and low order outlets.
- Use questionnaires to find out the sphere of influence of the different shops and services in each settlement. Other ways of measuring the sphere of influence include:
  - Questionnaires about where you have travelled from
  - Delivery or service areas of furniture shops
  - Distribution for a local newspaper or the places covered by the news in the paper
  - Client area for a vet, doctor or dentist
  - Membership area of a sports club
  - School catchments.

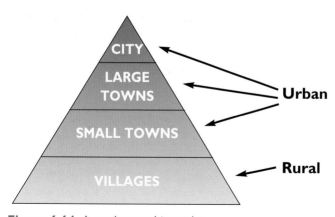

Figure 6.64 A settlement hierarchy

Shopping hierarchies

Figure 6.65 shows a shopping hierarchy. Notice that it is difficult to fit out-of-town shopping centres into the hierarchy because they are many different sizes and types, ranging from giant shopping centres around major cities to individual shops. Figure 6.66 shows the data you would need to collect.

Figure 6.65 A shopping hierarchy

Transport

Transport is rarely out of the news – increasing amounts of traffic on the roads, motorway congestion, gridlocked city centres, not enough parking spaces, problems on the railways, growing numbers of people travelling by air, airport expansion and increasing levels of noise for nearby residents. There are plenty of issues for investigation. Choose one that is of local importance if coursework on transport interests you.

What can you do?

1 In the CBD

- Where are the traffic bottlenecks – why are they there?
- Parking investigation – location and availability of car parks, use and adequacy, possible improvements and solutions.
- What are the advantages and disadvantages of pedestrianising streets in the centre?

Tackling Geography Coursework

1	Investigating size of settlements/shopping areas and the goods and services available	☐	Select a variety of different sized shopping centres that represent different elements of the hierarchy shown in Figure 6.65. Survey the different shops and services at each one.
		☐	If your main shopping centre is very large you could select a few shops to survey called indicator functions.
		☐	It might be a good idea to draw plans of each shopping area. You could also take some photographs.
2	Investigating sphere of influence and shopping habits – in general people will travel longer distances to buy comparison goods and do so less frequently	☐	For the questionnaire work, you could either ask a sample of people at each shopping centre or conduct a house to house survey on shopping habits. You need to ask how far people travel to use a centre, what they buy, frequency of use and mode of transport. Aim to do at least 50 questionnaires at each place, keep them short and simple. When asking about where they live, if it is a small shopping centre the street or local area is best, in a larger shopping centre the village, town or area.
3	Other investigations	☐	You could investigate the environmental quality, street appearance and shopping quality – this may be highest in the CBD and lowest at inner city corner shops.
		☐	Count pedestrians to find out the most popular service centres.
		☐	Count available car parking spaces and car parks.

Figure 6.66 Investigating shopping hierarchies – data collection

2 In the suburbs and surrounding rural areas

- Is a by-pass needed?
- How and why are traffic levels between various roads different?
- How much do numbers and types of vehicles along Road X vary at different times of the day?

3 In your home area

- How busy are the roads at different times of the day?
- What measures have been taken to manage traffic along the roads through the housing area and how successful have they been?
- Journey to work investigation – where do people on the housing estate work and what means of transport do they use?
- How much use do people living in your area make of public transport? Explain the results.

4 Near to a station or airport

- Investigate the impact (good and bad) of the bus station/railway station/airport on nearby residents.
- What is the sphere of influence of the station/airport?

- Hypothesis – people living close to the bus/railway station are more likely to use public transport than those living further away.

Methods of data collection

Questionnaires and traffic counts are the two methods used most in transport-related enquiries. Using questionnaires was dealt with in some detail on pages 24–27). When taking traffic counts, having a partner to help is often an advantage, especially when working in busy locations. The second person can either record traffic flowing in the other direction, or record movements by different types of vehicles. Planning where to do the counts is important, as also is deciding upon times and days of the week. Once you are ready:

- Record clearly location, starting time and finishing time.
- Use tally strokes on your recording sheet for each vehicle that passes.
- Place strokes in separate boxes for different types of vehicles.

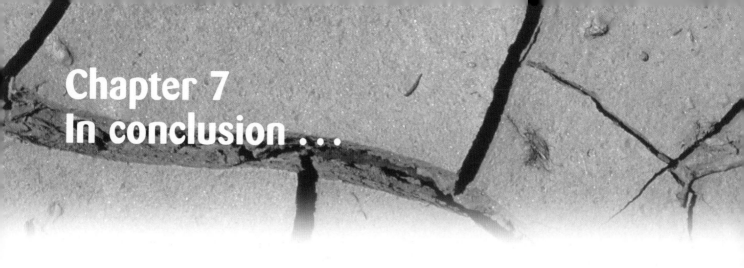

Chapter 7
In conclusion . . .

Levels marking

Many of you will already be familiar with levels marking. Levels of achievement will be used for marking your geography coursework in the examination.

Level 1 represents the lowest level of achievement. If all the marks awarded for your coursework are in Level 1, the best grade you can hope to achieve is Grade E. You are likely to do no better than Grade F or G. Some of the key features of work marked as Level 1 standard are listed in Figure 7.01.

Data collection

- Data collected and recorded in a group following precise instructions.
- No more than one or two methods of data collection used.
- Limited amount of data collected.
- Gaps and inaccuracies in data collected.
- Demonstrates only a simple understanding of what was being done and why.

Messages from this

- The student relies too much upon help from the teacher and other members of the group.
- The amount of data collected is less than was needed, and not all of it is accurate.
- Lack of understanding means that the student cannot explain where and why the data was collected.

Data presentation

- Limited number of methods of presentation used.
- Examples of incompleteness, for example, absence of titles and labels.
- Completed only after being given step-by-step instructions from the teacher.

Messages from this

- It is likely that only two or three techniques are used, typically bar graphs, pie graphs and photocopied maps.
- Many students at this level enclose maps, photographs and graphs without keys, titles and labels, or any attempt to relate them to the work.
- Despite over-reliance upon teacher help, some of the presentation remains inaccurate and incomplete.

Data analysis and conclusions

- Mainly a descriptive commentary about the data.
- Statements are brief and not explained.
- No real attempt to finish the work by relating back to the aims.
- Limited understanding and weak application.

Messages from this

- The written part is short and weakly related to title and aims.
- Description of data is regarded as a lower order skill than explanation.
- There is a huge gap in standards between this level and higher levels of achievement.

Figure 7.01 Some characteristics of coursework marked at Level 1 (the lowest level)

Level 3 is usually the highest level of achievement in geography examinations. Students hoping to achieve Grades A and B must achieve some marks within this band. Students need to produce work near to the top of the band in order to reach Grade A* standard.

Data collection

- Aims of study are explained and placed in their geographical setting.
- Initiative shown in making decisions about data collection (methods and locations).
- Work of data collection successfully carried out using a range of techniques.
- Understanding of what was being done and why is shown throughout.

Messages from this

- The coursework is properly set up with aims explained.
- Individual input is indicated by the reference to initiative.
- Accurate data collection, adequate in amount, is suggested.
- Strong understanding is demonstrated by the amount of explanation.

Data presentation

- Use of a good range of techniques.
- Including examples of those more complex in nature.
- Techniques used accurately and adeptly.

Messages from this

- The presentation adds to the work because of its variety and accuracy.
- Use of more complex methods demonstrates high geographical ability.
- It is likely that the student is in a strong position to analyse the data.

Data analysis and conclusions

- What is written matches the aims of the investigation.
- Short summaries throughout are followed by an overall conclusion at the end.
- There is a well ordered sequence to the work.
- Good detail is effectively used in line with title needs.

Messages from this

- All that is written is relevant to the title and aims.
- Key points are summarised along the way.
- The final conclusion completes the full investigation.
- It reads well because of its good organisation.

Figure 7.02 Some characteristics of coursework marked at Level 3 (the highest level)

TASKS

1 Look back at the extracts of the student's work on rivers (pages 70–73).

 Identify and describe the Level 3 characteristics for (a) data collection (page 71) and (b) data presentation (pages 72–73).

2 Look back at the examples of Student A's work on page 60. Why did none of this student's work reach Level 3 standard?

A coursework checklist

The following list shows you what the people who are going to mark and moderate your geography coursework are looking for. To achieve the highest marks, make sure that you cover these points in your work.

Introduction and Data collection

- **A precise title** ☐

- **A clear statement of aims in your *Introduction*** ☐

- **A brief guide to your area of study** ☐

- **Short statements about the wider geographical context of your work** ☐

- **Details about methods of data collection used** ☐

- **Explanations of where, when and why these methods were used** ☐

- **Data collection that was:**

  - **Accurate** ☐

  - **In line with the stated title and aims** ☐

  - **Sufficient** ☐

- **Recognisable individual input from you in the selection and collection of data** ☐

Data presentation

- **Use of a variety of data presentation techniques** ☐

- **Use of appropriate techniques for the types of data collected** ☐

- **Inclusion of one or two examples of more detailed or more complex techniques** ☐

- **Maps, tables, graphs, diagrams, sketches and photographs well placed within the text and labelled to support the aims of the study** ☐

Analysis and Conclusion

- **Key features of tables, graphs, etc., identified and described** ☐

- **What they show explained and analysed** ☐

- **Summary comments made along the way relating back to aims** ☐

- **Final *Conclusion* written in line with the main aim of the study** ☐

- **Short evaluation of strengths and weaknesses included** ☐

- **Mention of the wider geographical context of your work** ☐

Just before you think you have finished, complete the boxes in the checklist to see if you have missed anything out. This is another example of the benefits of using a word processor, because it is so easy to insert and make changes. Any changes at this stage should only be minor. They should be nothing more than occasional extra sentences, small tweaks and last-minute improvements such as:

- **Making the statement of aims in the *Introduction* even clearer** ☐

- **Changing the position of a map so that it is closer to the writing to which it refers** ☐

- **Adding an extra point to the evaluation.** ☐

TOP TIPS

- Never use plastic sheets and ring folders – moderators hate them.
- Number all the pages.
- Place all the A4 sheets loose in a flat folder with a flap.
- This is neatest way to present your work.
- It will show the moderator that you are well organised and in control.

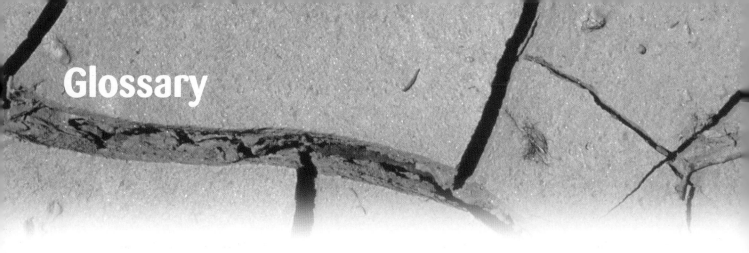

Glossary

Aim – What you intend to find out.

Analysis – Commenting on what the data shows, looking to identify the most important features and key patterns, and recognising their significance.

Annotate – Add labels or notes to maps, diagrams, photographs, etc., in order to highlight their main geographical features.

Bibliography – A list of secondary sources used while undertaking the coursework.

Conclusion – The final summary that takes into account findings from all the data collected.

Correlation – This tests the strength of a relationship between two sets of data, or whether the relationship exists.

Data – Geographical information that is collected by fieldwork or from *secondary sources*.

Distribution – Where features are located in an area or on a map. It may be possible to recognise patterns.

Environmental – How the physical features of the Earth have and are being affected by the presence of people on Earth.

Evaluation – Taking an overall look at your work and thinking about what was successful and what could have been done better. It shows the degree to which you are satisfied with your work.

Extension work – Extra data collection and presentation, undertaken by yourself, beyond what was done in teacher-led or group work.

Fieldwork – New data collection, usually undertaken out-of-doors.

Function – The main purpose.

Functional zone – An area which shares one main purpose, such as a housing zone or a zone of industry.

Geography – The study of places, concentrated on features of the Earth's surface, and the physical and human factors responsible for them.

Group work – Two or more people working together, usually to collect data.

Hierarchy – When settlements or shops are arranged in order of size and importance from lowest to highest in the shape of a pyramid.

Honeypot – An attraction or place that draws large numbers of visitors to it.

Hypothesis – A statement about what you expect to find before you undertake the actual investigation. A hypothesis can be tested to prove whether or not it is correct.

ICT – Information and Communications Technology, usually associated with the use of computers in a coursework context.

Individual initiative – Working alone so that you are responsible for all the planning and doing.

Interview – Asking questions of a person, usually on a one-to-one basis.

Issue – Something of concern to people, sometimes a conflict between groups of people with different views.

Land use – Different ways of using the Earth's surface; often rural land uses (farming, woodland, marsh, etc.) are distinguished from urban land uses (houses, factories, parks, etc.).

Measurement – Using items of equipment to find out height, depth, length, slope angles, etc.

Observation – Transferring what can be seen by the eyes onto paper, for example, by writing notes, drawing sketches and making assessments.

Pollution – Humans changing the natural environment so that it is spoiled in some way.

Presenting data – Displaying information in a more visual way, for example, in tables, graphs and diagrams.

Primary data – New *data* obtained by you by undertaking fieldwork; methods include *observation*, *measurement* and asking questions (*questionnaires* and *interviews*).

Processing data – Using raw data to calculate other values.

Quadrat – A square-shaped frame, within which surface features are studied such as amount and types of vegetation and pebble sizes.

Questionnaire – Sheet of questions for people in the survey to answer.

Random – Selection by chance; a totally haphazard selection of people or places, widely used in *sampling*.

Rank – Values placed in order from highest to lowest.

Rural settlement – Villages, hamlets and farms are examples of settlement types in country areas; only small areas of land are built up.

Sampling – Studying a number of people or places for data collection that is smaller than the total number, but which hopefully will give a good indication of the total picture.

Secondary data – Old *data* obtained by someone else, made available to you by having been published (for example, in books and newspapers), being available on the Internet or obtainable from organisations (for example, Local Councils and Tourist Boards).

Sketch – Hand-drawn summary; it can be a map or a drawing showing landscape features.

Sphere of influence – The area served by a settlement, shop or service. It is determined by the greatest distance that people travel to visit the settlement or attraction, or use the service.

Stratified sample – Choosing places or people for study, taking relative sizes into account.

Systematic sample – Choosing places or people in a regular manner (for example, every 10).

Title – What you are going to find out, the main aim.

Total population – When information about every single feature or all the people in an area can be obtained.

Transect – Following a line in order to collect fieldwork data. Lines often follow roads in studies of urban geography; straight lines are more likely to be used for taking measurements in physical geography.

Urban settlement – Towns and cities are large built-up areas, with their own shops and other services.

Index

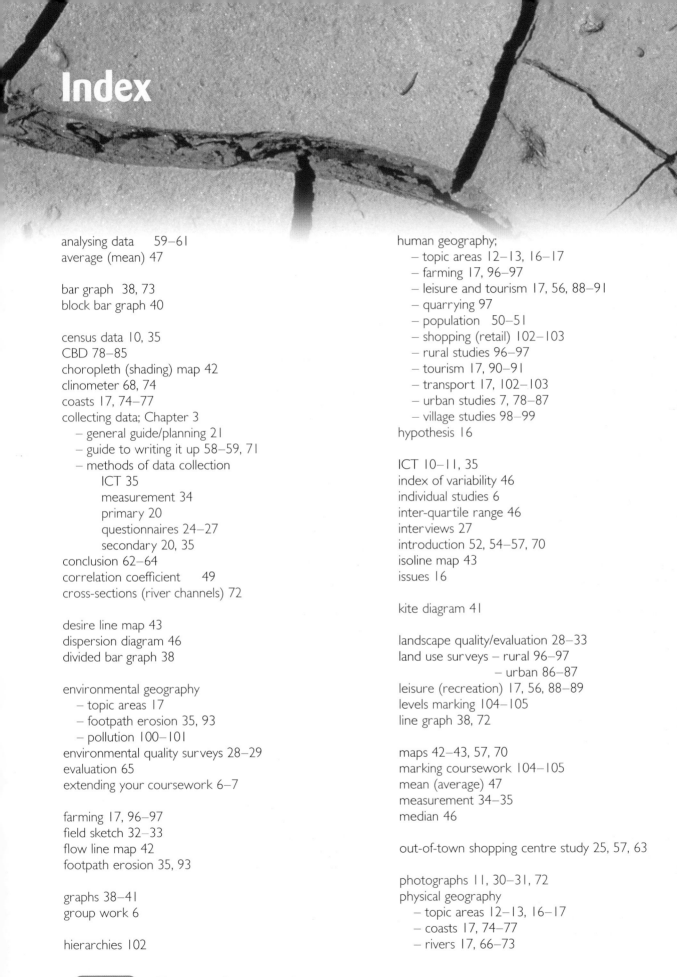